I0109620

Time's

Second

Arrow

Also by Robert M. Hazen

Symphony in C: Carbon and the Evolution of (Almost) Everything

The Story of Earth: The First 4.5 Billion Years, from Stardust to Living Planet

Genesis: The Scientific Quest for Life's Origins

The Diamond Makers

Why Aren't Black Holes Black? (with Maxine Singer)

Science Matters: Achieving Scientific Literacy (with James Trefil)

The New Alchemists: Breaking Through the Barriers of High Pressure

The Breakthrough: The Race for the Superconductor

Keepers of the Flame (with Margaret Hazen)

The Music Men (with Margaret Hazen)

Wealth Inexhaustible (with Margaret Hazen)

The Poetry of Geology

Time's Second Arrow

■

Evolution, Order, and a New Law of Nature

Robert M. Hazen
and *Michael L. Wong*

W. W. NORTON & COMPANY
Independent Publishers Since 1923

Printed in the United States of America
First Edition

For information about permission to reproduce selections from this book, write to Permissions, W. W. Norton & Company, Inc., 500 Fifth Avenue, New York, NY 10110

For information about special discounts for bulk purchases, please contact W. W. Norton Special Sales at specialsales@wwnorton.com or 800-233-4830

Manufacturing by Lakeside Book Company
Book design by Anna Knighton
Production manager: Gwen Cullen

ISBN: 978-1-324-10548-0

W. W. Norton & Company, Inc.
500 Fifth Avenue, New York, NY 10110
www.wwnorton.com

W. W. Norton & Company Ltd.
15 Carlisle Street, London W1D 3BS

Authorized EU representative: EAS,
Mustamäe tee 50, 10621 Tallinn, Estonia

10 9 8 7 6 5 4 3 2 1

for

Anirudh,

Carol,

DJ,

Heather,

Jim,

Jonathan,

and

Stuart

If you wish to make an apple pie from scratch,
You must first invent the universe.

—Carl Sagan, *Cosmos*

Contents

Preface

What is this mysterious, fleeting phenomenon that we experience as "now"?

And now?

And now?

For better or worse, we remember the past, but we can never remember what is yet to be.

No one escapes time's arrow. An incessant drive toward disorder and decay fills our lives. In its most quotidian manifestations, it dictates that our new shoes eventually become scuffed, our new car dinged. Likewise, the arrow of time points our fragile bodies on an inexorable path of decline, death, and decay. Moments that have passed are irreversible. That terrible thing you said—the one you wish could be unsaid—is beyond unsaying.

This idea—that the universe trends toward randomness and chaos—is so powerful that it is enshrined in one of the foundational laws of nature. Indeed, in the canon of existing natural laws only one statement—the second law of thermodynamics—embeds a direction to time, and it declares that "*the disorder of a closed system must increase.*" Entropy will have its way.

Yet in your heart and mind, you know that such a dispiriting framing cannot be the complete story of nature's ever-changing face. After all, the universe started almost fourteen billion years ago with the Big Bang—a time when there were no atoms or molecules, no planets or stars, no people nor any books for them to read.

Look around you today at the pervasive wonders of our world: Flowers bloom, birds sing, rainbows grace the luminous sky. We watch in awe as our children are born, grow, and learn. From wordless beginnings, humans have created poetry, art, music, and science. Everywhere we look, miraculous *order* emerges and novel patterns form.

These shared human experiences reveal an obvious truth: There must be a second arrow of time. There must be a missing law of nature—one that speaks to how the universe trends toward such remarkable states of intricate organization and flamboyant behavior.

How can it be that such an omnipresent, awe-inspiring facet of the cosmos is uncodified in the canon of scientific law?

▪

It is often said that scientists base their conclusions on reproducible, independently verifiable evidence wherever it might lead, not on preconceptions or beliefs. The reality of atoms, the existence of black holes, and the wonders of biological evolution are not matters of faith. These and countless other scientific concepts arose from sound, logical conclusions based on careful observations and measurements of the natural world. What's more, as tested and retested as such concepts as atoms, black holes, or evolution might be, if new reproducible data indicate a more plausible model, then even the most established scientific ideas will be modified or ultimately discarded. In this context of skeptical inquiry, an open, questioning mind is the scientist's most valuable character trait.

That view of the detached, objective scientist, unfettered by any biases or beliefs, is comforting to those of us who study nature for a living. But it's not entirely true. Scientists do have a few unshakable core beliefs. We fervently believe that *nature obeys laws*. Furthermore, we are confident in our belief that the scientific enterprise is the best way, perhaps the only way, to discover and describe those laws. And, for those many scientists who have faith in the existence of a Creator, the quest for nature's laws—the pure pursuit of understanding how and why creation works the way it does—is the surest path to knowing the mind of God.

A core objective of science is to codify every natural phenomenon on Earth and in the heavens in a concise explanatory and predictive framework. We experience innumerable events every day—a perpetual cacophony of sights and sounds, forces and motions in a cosmos where change seems the only constant. We strive to learn if there are underlying patterns that might provide a sense of order to that seeming chaos. In that context, a crowning scientific achievement of the past five hundred years has been the discovery of just such a framework—the articulation of a set of natural laws that, taken together, confer remarkable order on the seeming randomness of our experiences in the physical universe.

In this book we propose that at least one new law of nature—a law that describes and explains the increase in order that has been manifest by myriad complexly patterned and evolving systems from the primordial Big Bang to today—must be added to the list. This law, we argue, is a law of *universal* evolution—not only in biological organisms but also in the widest range of complex systems. Our new law applies to the evolution of language, art, and music; to advances in technology, artificial intelligence, and scientific knowledge; and, surprisingly, to diverse nonliving systems such as atoms, molecules, minerals, and stars.

If our theory is correct, then the implications for all of us are profound. A valid description and explanation of how our cosmos becomes ever more ordered and complex might provide a road map for how we, both as individuals and as a society writ large, can exert agency on future change. Not only will a universal law of evolution offer a new understanding of nature's constantly changing face, but it may also provide new strategies to exercise control on potentially unruly evolving systems, from the global climate to cancer cells to artificial intelligence.

Our exploration is just beginning, and we invite you to join us in this journey of discovery.

Time's Second Arrow

1

The Laws of Nature

We live in a cosmos of impartial, immutable, universal order. Every action has a cause. Every desire, every ambition is constrained by natural law.

Nature's laws exist at varied scales of space and time, with many phenomena lying far beyond our five senses. We do not directly experience the counterintuitive, atomic-scale quantum world. Travelers of our generations will never encounter the strange behaviors of objects moving close to the speed of light. Similarly, the relativistic space-time distortions of masses large enough to create black holes are far beyond anything we can imagine. Modern physics has made astonishing technological progress in extending our senses to probe such unfamiliar realms. Patterns of strange, yet lawful, behavior continue to emerge from such investigations, but these extreme environments play almost no perceptible role in our day-to-day lives.

Rather, our everyday concerns are addressed by the macroscopic natural laws—the behaviors of matter, energy, forces, and motions at the familiar scales of time and space that frame human experience. Why do objects move? How does gravity hold us to Earth? What are heat and light? When did life emerge, and how has it evolved? These are the questions that affect us every day of our lives. It is little wonder, then, that we so desperately seek universal, predictive laws that describe and explain what we experience firsthand—the natural laws that dictate what we can and cannot do in a physical universe.

Science has responded with a remarkable suite of macroscopic laws of nature—a dozen statements, give or take a few, that codify almost everything we can and do experience in the tangible world. Each law of nature describes and explains a constellation of related phenomena—the systematic way that forces influence objects in motion, for example, or limitations on the ways that energy can shift from one form to another. These transformative macroscopic laws of nature go far beyond explanations of what has happened. They enable sound predictions of what might be, as well as sobering reminders of what can never be. Thanks to the laws of nature, every facet of our evolving technological world rests on a firm theoretical foundation. Each law is a statement with profound consequences:

THE MACROSCOPIC NATURAL LAWS

Newton's Laws of Motion

1. *First Law of Motion*: A body will continue in its state of rest, or of uniform motion in a straight line, unless it is compelled to change that state by forces imposed upon it.

2. *Second Law of Motion*: Force equals mass times acceleration.

3. *Third Law of Motion*: To every action force there is an equal and opposite reaction force.

Newton's Law of Gravitational Attraction

4. Between any two objects there is an attractive force proportional to the products of the two masses divided by the square of the distance between them.

Laws of Electromagnetism

5. *Magnetostatics*: Every magnet has two poles; like magnetic poles repel each other while unlike poles attract.

6. *Electrostatics (Coulomb's Law)*: The force between any two electrically charged objects is proportional to the product of their charges divided by the square of the distance between them.

7. *Electromagnetism*: Magnetic fields are created by moving electrical charges.

8. *Electromagnetic Induction*: Electrical fields and electrical currents are created by changing magnetic fields.

Laws of Thermodynamics

9. *First Law of Thermodynamics*: In a closed system, energy can transform from one form to another many times, but the total amount of energy is conserved.

10. *Second Law of Thermodynamics*: Heat will not flow spontaneously from a colder to a warmer body.

Add to these ten statements a few boundary conditions—both mass and electrical charge are conserved, for example—and you have a remarkably robust description of the tangible cosmos that we experience.

How were these laws discovered? What do they mean to us in the modern world? And can we be sure that this list is complete—that we haven't missed something important?

Isaac Newton's Three Laws of Motion

Forces and motions pervade every moment of our lives. Our beating hearts pump blood through arteries and veins. Our lungs force oxygen in and carbon dioxide out. Our legs propel us across uneven ground. Without the lawful behavior of forces and motions, our technologies would fail—no landings on the Moon, no self-driving cars, no reliable machines of any kind.

There was a time not so long ago when humans were skeptical that laws of nature existed at all—a time when the cosmos seemed uncaring and arbitrary in dealing out disease and death. The devastating consequences of droughts, floods, fires, and famines were capricious and unpredictable. When a skilled hunter missed their aim or a prized pot fell and shattered, it was often easier to blame the whims of the gods than admit to human error.

The ancients hedged their bets. The planetary models of astronomers in India and China, the observatories of the Maya and Babylonians, and the enduring monuments of Stonehenge in England and Newgrange in Ireland testify to the conviction, established thousands of years ago, that at least some aspects of the natural world are predictable. In spite of life's uncertainties, the rising and setting of the Sun, the phases of the Moon, the motions of the planets, and the endless cycle of seasons provided a sense of cosmic stability—reliable patterns to mark the inexorable passing of the days, months, and years. Such predictions were not of merely abstract interest to societies struggling to survive. Knowing when in the annual cycle to plant and when to harvest are keys to successful agriculture. Hunting game during the full Moon, or attacking your enemy at the new Moon, were essential strategies for survival. But a deeper understanding of how and why motions occur in the heavens or on Earth remained elusive.

Despite pervasive superstitions and magical thinking, by the 1600s a kind of municipal pragmatism gradually prevailed. Knowledge is power. At a time when rulers often looked to astrology and the position of celestial bodies to guide their policies, understanding the predictable motions of planets and moons was imperative. In an age when gunpowder and cannon were keys to seizing and retaining power, the lawful flight paths of cannonballs were of supreme importance.

Religious precepts created some thorny roadblocks to discovery. In particular, idealized laws of heavenly motion ("celestial mechanics") were assumed to differ in fundamental ways from those of the corrupt earthly realm ("terrestrial mechanics"). Earth, by church doctrine, was situated as the fixed cosmic center, the fires of hell just a short commute beneath our feet. Yet progress on the nature of forces and motions, once the right questions were posed and tackled, was inevitable. Laws of forces and motion are there for all to see, waiting to be discovered by the curious mind.

Many scholars, dating back to Aristotle and before, thought about the origins of motion. Their strategy was to apply pure logic in preference to observation to understand nature. Galileo Galilei (1564–1642), the famed Florentine physicist, represented a new generation of natural philosophers (what we now call scientists) who championed a powerful alternate approach—the experiment. Supported with government funding by the ruling Medici family, Galileo plotted the parabolic trajectories of flying objects—empirical research that significantly improved the accuracy of artillery. Other experiments documented the acceleration of balls rolling down an inclined ramp and the attributes of swinging pendulums, providing hints of a deeper set of predictive lawful processes. These advances, with their myriad practical applications, gained Galileo acclaim.

With his revolutionary astronomical research, Galileo ventured into more dangerous intellectual ground. The first to study the heavens using a telescope—a device Galileo did not invent, but one he substantially improved—he clearly saw what no one had seen before. He discovered four moons orbiting Jupiter (which he strategically named the Medicean stars in honor of his patron). He discerned phases of Venus like those of the Moon. He documented craters on the Moon, as well as spots on the Sun and an odd distortion of Saturn (what we now recognize as rings). His observations also clearly implied that all planets, including Earth, orbit the Sun.

The Catholic hierarchy was not amused. Earth must be at the center of the universe. Celestial objects are perfect spheres. Moons orbiting Jupiter or phases of Venus violated doctrine, as did blemishes such as craters on the Moon. The legend of Galileo resisting claims of heresy, and then recanting when shown the instruments of torture, is well known. His scientific story ended with a decade of study and thought while under house arrest—the time during which he refined his most influential research on the uncontroversial behavior of moving objects at Earth's surface. Galileo Galilei never articulated any of the macroscopic laws of nature, but by championing a more empirical approach to science, wherein new observations could challenge old perceptions, he laid the groundwork for the next generation of scholars.

Enter the young Isaac Newton (1642–1727), born in the year of Galileo's death in the pre-Cromwellian England of Charles I. Newton's was not an easy childhood: He was predeceased by his father, abandoned by his mother at age three, sent to boarding school at twelve, and reunited with his mother at age seventeen only to be forced to run the family farm—a profession he despised. Newton was "saved" from that agrarian life by his academic brilliance. A child prodigy in mathematics and invention, he was admitted to the University of Cambridge's Trinity College in 1661. Newton, though by no means a standout student at Cambridge, immersed himself in the literature of science and mathematics—works of Aristotle, René Descartes, Galileo, and other leading authorities. He had found his calling in the life of the mind.

The Great Plague of 1665–1666, which would eventually kill one in every seven Londoners, all but shut down England. Cambridge University closed, sending students home for almost two years. Newton, then in his early twenties, spent a year and a half on his family farm in the rural Lincolnshire hamlet of Woolsthorpe-by-Colsterworth, studying and thinking on his own. There followed one of the most remarkable sequences of discoveries in the history of science. In that short, intense interval, Isaac Newton made several revolutionary advances: He invented integral calculus, set down the laws of optics, and articulated four macroscopic laws of nature—three universal laws of motion and the universal law of gravity. His discoveries were soon circulated among scholars in manuscript form, though it was not until 1687 that these and other findings were collected into the three-volume treatise *Philosophiæ Naturalis Principia Mathematica*, now universally known as the *Principia*.

Newton's development of the laws of motion required a striking conceptual advance. Natural philosophers distinguished two kinds of motion, uniform and accelerating. Prior scholars thought of uniform motion as including both objects traveling in a straight line at constant speed (a frictionless ball rolling on a flat table, for example) and objects traveling in a perfect circle at constant speed (presumably the situation of celestial bodies in orbit around Earth). Everything else was nonuniform motion, or acceleration.

Newton saw motions differently. To him, uniform motion only applied to objects that are stationary or moving in a straight line at constant speed. Every other kind of motion—slowing down, speeding up, going around a bend, or traveling in a perfect circle about Earth—was acceleration. All those examples involve a *change* from uniform motion. In Newton's view, the motions of a slowing-down carriage, a speeding-up horse, or an orbiting planet are conceptually equivalent. All are unambiguous examples of acceleration. With this breakthrough, the universal laws of motion followed.

The First Law of Motion: The first law of motion introduces Newton's definition of "force."

A moving object will continue moving in a straight line at a constant speed, and a stationary object will remain at rest, unless acted on by an unbalanced force.

In one sense, the first law of motion seems obvious, almost trivial. If left alone, an object won't change its state of motion. Conversely, if you throw it, pull it, or otherwise disrupt its motion, then it will accelerate. But Newton adds that bit at the end about an "unbalanced force"—implying that there are natural phenomena called forces and that they can be applied to accelerate things. The first law of motion says nothing about what those forces might be, or how many kinds of forces occur in nature. But armed with the first law, one can start the search.

Newton's idea was radical. As we've seen, most prior scientists from the times of Aristotle to Galileo would have argued that the first law is wrong. The circle is the most perfect geometric shape, so in heaven objects must move in perfect circles *without any forces involved*. Newton countered with a familiar example. Tether a ball to a string and swing it around your head in a circle. Let go and the ball doesn't keep going in a circle—it flies off in a straight line.

Newton's first law tells us something profound. When we see a change in uniform motion—an acceleration—then a force must have acted to produce that change. In this way, the first law of motion alerts us to when a force must be acting.

The Second Law of Motion: The second law of motion is an equation that quantifies how forces and acceleration are intertwined, while introducing the critical concept of "mass."

> The acceleration produced on a body by a force is proportional to the magnitude of the force and inversely proportional to the mass of the object.

Newton's first law of motion tells us when a force is acting. The second law of motion tells us exactly how much acceleration to expect when that force is applied: Force equals mass times acceleration. This equation depends on the physical quantity called mass—a measure of the amount of "stuff" an object contains. On Earth we can compare the relative masses of different objects by weighing them, but mass is not equivalent to weight. Even in the weightlessness of space, an object has the same mass as on Earth, and that mass will require a force to accelerate it.

Newton's second law is immensely practical. It states that if you want greater acceleration, you need a larger force, and the larger the mass, the greater the force required to achieve a given acceleration. Your everyday experience confirms this intuitive law: It's easier to lift a bicycle than a car. It's easier to throw a golf ball than a bowling ball.

The power of the second law of motion lies in its predictive ability—a benefit that permeates our technological society. All manner of accelerations, from the press of a button to the launch of a spaceship, are dictated by this law. Acceleration is a balance between force and mass. Whether running a sprint or maneuvering a Moon rocket, if you want to achieve greater acceleration, then you must either increase the force or decrease the mass. Thus, while Newton's first law defines the concept of force as something that causes a mass to accelerate, the second law goes much further by defining the magnitude of the force necessary to achieve a given acceleration of a given mass.

The Third Law of Motion: Newton's third law of motion is at once one of the most familiar yet counterintuitive ideas in science. Newton said:

For every action there is an equal and opposite reaction.

Newton's third law of motion can be recast in a simpler form: *Forces always act in pairs.* The key idea embedded in this law is that whenever a force is applied to an object, that object must simultaneously exert an equal and opposite force. Think about it. When you throw a ball, you accelerate the ball and it flies off from your hand. But while you accelerate the ball, the ball is pushing back with an equal and opposite force on your hand. Push a button and the button pushes back on your finger. Wind pushes on a sailboat, propelling it across the water, but simultaneously the sail pushes against the wind, causing it to shift in direction and speed.

Sometimes the pairs of force are less intuitive. We naturally think of our experiences in terms of causes and effects, often where bigger, faster objects exert forces on smaller, slower ones: A golf club drives a golf ball, a car windshield hits a bug, a boxer punches her opponent's nose (ow!). But in terms of Newton's third law, all these events must also happen simultaneously the other way around. The golf ball slightly alters the club's swing. The bug exerts an equal and opposite force, causing an imperceptible deceleration of the car, even as the bug is squashed. The boxer's nose changes the direction and speed of her opponent's punch. (Though still—ow!)

Newton's three laws of motion, always interdependent and intertwined, pervade every action we take, every moment of our lives. Even the most quotidian of daily events must exemplify these universal laws. Pick up a cup of coffee. You have caused an acceleration and thus must have applied a force (first law). You intuitively balanced the force you applied to the mass of the coffee cup to achieve a relaxed acceleration (second law). The cup exerted an equal and opposite force on your hand (third law).

Newton's laws of motion, a contribution of pure genius, were enough to guarantee him a place in the pantheon of science greats. But he wasn't done. He next turned his attention to one of the fundamental forces in nature—the invisible, pervasive force of gravity.

The Law of Gravity

That gravity is a force on Earth was not a mystery to Newton or his contemporaries. After all, things fall. Galileo's experiments—rolling balls, pendulums, cannonballs—had shown the effects of gravity in detail. Objects falling near Earth's surface, he observed, accelerate at a constant rate, increasing their speed by roughly thirty feet per second every second of fall. Nothing accelerates without a force (the first law of motion), so gravity must be one of nature's forces.

The difficulty in devising a *universal* law of gravity lay in comprehending the heavens. The old view was that celestial objects travel in perfect circles—a presumed kind of uniform motion that required no forces at all. But Newton's first law clearly states that circular orbits require a sustaining force. Otherwise, orbiting planets and moons would fly off in straight lines.

By Newton's own account, the moment of inspiration came as he was sequestered on his family farm, sitting under an apple tree. He looked up and saw ripe apples on the tree, with the

Moon in the distance. (Counter to some amusing variants of the story, an apple did not hit him in the head. Well, maybe it did metaphorically.) Newton knew that an apple would fall downward from the tree. Why wasn't the Moon also falling?

In fact, he realized, the Moon *is* falling all the time. Think of it this way. If you pick up an apple and throw it sideways, it travels in an elegant arc—a parabola, as Galileo had demonstrated for cannonballs. Throw the apple harder and it flies farther in a wider parabolic arc before falling to the ground. Now imagine being up in space and flinging an apple so hard that it simply keeps falling around the "edge" of the planet. The apple, like the Moon, would "fall" forever in an orbit. Newton made the intuitive leap that others had missed: the apple and the Moon are both falling. Gravity is a universal force.

Moreover, as the German mathematician and astronomer Johannes Kepler (1571–1630) had shown more than a generation earlier, the orbits of planets and moons are elliptical—not perfect circles. The parabolic paths of thrown objects on Earth and the elliptical paths of planets in orbit are mathematically similar in character (both are described by so-called quadratic equations), suggesting to Newton a simple mathematical form for a universal natural law:

Between any two objects in the universe there is an attractive force (gravity) that is proportional to the masses of the objects and inversely proportional to the square of the distance between them.

Let's deconstruct that mouthful of a natural law. Newton is telling us that an attractive force—the gravitational force—

exists between any pair of objects with mass. Those two objects could be Earth and the Moon, a saltshaker sitting next to a pepper grinder, or two molecules in the atmosphere. The magnitude of the gravitational force is proportional to each of the two masses—greater masses mean proportionally larger attractive forces.

The other critical factor in determining the magnitude of the gravitational force is the distance between the two objects. Proximity is important. Double the distance between the two objects and the force is reduced to a quarter of the original value. Triple the distance and gravity is only one-ninth as strong. But halve the distance between the two masses and gravity increases by a factor of four. That kind of distance relationship is called an inverse square law—a pattern seen in many natural phenomena from electrostatics to stage lighting.

The predictive power of Newton's universal law of gravity was put to the test in 1682, when British astronomer Edmond Halley (1656–1742) carefully measured the elliptical path of a bright comet (now known as Halley's Comet). He used Newton's law of gravity to compute the orbit and to predict its return in 1758. Halley's Comet next appeared right on schedule, on Christmas Eve of 1758—an event celebrated worldwide as a triumph for Newton's law of nature.

The Laws of Electromagnetism

Gravity can't be the only force in nature. After all, airplanes fly up into the sky, lava fountains erupt hundreds of feet above volcanic peaks, and leaves blow hither and yon in the wind. Trees grow tall, rain-filled clouds float across the sky, and we can walk up a flight of stairs without a second thought—all in the context of Earth's incessant gravitational pull. But what other forces might there be, and how many different forces play a role in our daily lives?

The surprising answer is that we experience only one other force—the electromagnetic force. The sight of a storm cloud, the roar of thunder, the smell of petrichor, the tickle of a raindrop on your nose—all are expressions of the electromagnetic force at work.

Unlike Newton's laws of motion and gravity, the laws of electromagnetism did not emerge fully formed from a single mind. More than two centuries of study by a score of researchers, often focused on topics that appeared to have no connections whatsoever, culminated in a set of four universal natural laws—the laws of electromagnetism.

The Law of Magnetostatics: Long before Newton's discoveries of the laws governing forces and motions, practical mariners exploited the mysterious force of magnetism to guide their journeys across the seas. More than two thousand years ago, Chinese scholars discovered the surprising properties of the naturally magnetic mineral magnetite, which attracts other bits of magnetite as well as iron metal. Elongated pieces of magnetite were soon fashioned to make the earliest known compass needles, which rotate in a plane until they point to Earth's north magnetic pole.

A thousand years later, Chinese compass makers exploited the strange ability of magnetite to transfer its attractive power to a piece of iron that is stroked repeatedly in one direction. Magnetized iron needles carefully balanced on a pin became the norm for compasses, which benefited European navigators as early as the fourteenth century.

But how does a compass work? Why does the compass needle

move? This mysterious phenomenon, implying the influence of an invisible force, demanded careful study. English physician and natural philosopher William Gilbert (1544–1603), perhaps best known in his day as Queen Elizabeth's doctor (he was at her bedside when she died), made several key discoveries, combining prior scholars' observations with his own meticulous experiments. Breaking magnets into smaller and smaller pieces, Gilbert realized that every magnet, however small, has two poles, one that points north and the other south. Furthermore, he found that north and south poles attract each other, while like poles repel—measurements that would play an important role in the formulation of Newton's laws of motion almost half a century later. Gilbert also correctly speculated that Earth itself is a giant magnet with two poles, possibly with an iron core.

Taken together, Gilbert's research suggests a simple, two-part law describing the universal behavior of magnets—a law of magnetostatics:

Every magnet has two poles; like magnetic poles repel each other, while unlike poles attract.

In his influential treatise, *De Magnete*, published in scholarly Latin in 1600, Gilbert presented both his varied experimental results and his philosophy regarding the primacy of conducting experiments to understand the natural world. To this end, his research extended to other natural phenomena, including electrostatics—the properties of objects influenced by static electricity. His conclusion—that electricity and magnetism are two distinct phenomena—would prevail for the next two centuries.

The Law of Electrostatics (Coulomb's Law): Experiences of a dry winter's day—static cling, frizzy hair, clingy plastics, and little electric shocks after walking across a wool carpet—present an explanatory challenge in the context of the law of magnetostatics alone. By what mechanism do these nonmagnetic objects stick together, push apart, and deliver unwelcome jolts? Natural philosophers, including William Gilbert and his predecessors, devoted considerable attention to this pervasive, if seemingly inconsequential, phenomenon. Little could they imagine how electricity might someday change the world.

In eighteenth-century Europe and America, a group of scholars followed in Gilbert's footsteps, probing the nature of electrical effects. They called themselves "electricians," a name derived from "electron," the Greek word for amber, which had been employed since ancient times to produce a static charge by rubbing against a piece of fur. Objects thus charged repelled each other with static electricity (so called because the electric charge doesn't move in these experiments). A similar repulsive effect was produced between objects charged by rubbing a sheet of glass with silk. But if you brought an object charged by amber next to one charged by glass, the two would be attracted to each other.

What was going on? It was Benjamin Franklin (1706–1790), the American patriot and signer of the Declaration of Independence, who proposed that a single "electrical fluid" (what we now call electrons) could explain all these phenomena. In Franklin's model, negatively charged electrons are usually in balance with their positively charged hosts. There is no net electrical charge. But certain materials when rubbed together will lose or gain electrons. Fur rubbing against amber or silk rubbing against glass creates an imbalance, a static electric charge. The charge can be either negative, if the object has extra negative electrons, or positive, if the object has an electron deficit. The electric force arises because opposite charges attract, while like charges repel.

Franklin's discovery of an electrical fluid had an immediate practical benefit. He suspected that lightning, a dangerous and destructive phenomenon in a time of wooden buildings, was caused by the large-scale buildup of electrical fluid in clouds. This fluid, he knew, easily flowed through metal. His invention of the lightning rod, a simple metal wire projecting above the roofline and leading into the ground, saved countless lives.

Franklin's research came close to exposing a natural law of static electricity, but it lacked the quantitative rigor required for a truly predictive law. French physicist Charles Augustin de Coulomb (1736–1806) made the key discovery in 1785, about the same time that Franklin and his American colleagues were inventing the United States Constitution in Philadelphia. Coulomb devised an experimental apparatus that allowed him to place carefully controlled positive and negative charges on objects and then measure the attractive or repulsive forces between them. The result was a quantitative law of electrostatics, now known as Coulomb's law:

The force between any two electrically charged objects is proportional to the product of their charges divided by the square of the distance between them.

This statement might seem familiar. Indeed, it is almost identical in form to Newton's universal law of gravity. Just replace "two electrically charged objects" with "two masses." Of course, the gravitational and electrostatic forces are different in critical ways. One relates to mass, the other to electrical charge. Gravity is always attractive, while the electrostatic force can be attractive or repulsive. Nevertheless, the similarities provide a reassuring sense that the cosmos possesses a kind of satisfying, if well hidden, order. Yet none of the early pioneers in studies of magnetism and electricity could have imagined the levels of hidden order yet to be discovered in—of all things—frogs.

The Law of Electromagnetism: Frogs? What could frogs possibly have to do with discovering the laws of nature? The story—one of the most surprising twists in the history of science—reveals how research in one domain may profoundly influence another.

Italian biologist and physicist Luigi Galvani (1737–1798), a professor at the University of Bologna, tackled a knotty problem—"animal electricity," what was thought by some to be the motive force in the biological realm. Galvani focused on the behavior of frog legs, which when dissected to expose muscles and nerves and then subjected to electrical shocks, would kick and twitch long after death. On one occasion, Galvani inadvertently touched a frog nerve with metal forceps. Even though no spark was applied, the leg began to twitch. Galvani was convinced that he had tapped into the animating electrical force of life itself—an idea amplified by Mary Shelley's dystopic vision of Dr. Frankenstein reanimating the dead with electricity.

Alessandro Volta (1745–1827), a physics professor at the University of Pavia, disagreed. The behavior of the frog legs, he argued, was the result of Galvani's experimental design. An electrical effect arose when the forceps made of brass touched the iron metal hook from which the legs were suspended. The resulting electrical circuit spontaneously generated electricity.

After much debate and competitive jousting—the good citizens of Pavia and Bologna took sides, much as in a football match between collegiate rivals—Volta settled the matter for good by inventing the battery in 1799.

Volta's first batteries, also called Voltaic piles, were simple, clunky affairs. He filled a wooden box with alternating zinc and silver plates, separated by cloth layers and filled with salt water. Wires protruded from each end of the box. In such an arrangement, electrons spontaneously flow from zinc to silver. Armed with these batteries, electricians could for the first time experiment with a continuous flow of electrons—an electric current rather than episodic sparks. Electric circuits soon followed, with special focus on chemical experiments. Place the two battery wires in water, for example, and pure oxygen gas will bubble up from one end, while twice as much pure hydrogen will bubble up from the other. More electrical discoveries followed during the first decade of the nineteenth century. Nevertheless, any connection between electricity and magnetism remained elusive.

The mythology of science history recounts that a Danish professor of physics and chemistry, Hans Christian Oersted (1777–1851), made the breakthrough in 1820 during a classroom lecture on electricity and magnetism. Oersted, so the story goes, used a battery to pass electrical current through a coil of wire when he noticed a nearby compass needle simultaneously twitch and turn. In fact, he had been conducting similar experiments since 1818, but only gradually realized that the coil of wire created a dipole magnetic field, exactly like that of an iron magnet. He had invented the electromagnet—a magnet that you can turn on and off.

A flood of practical inventions, including magnetic switches, the telegraph, and the electric motor, followed Oersted's revelation. So, too, did the first statement of a natural law that linked electricity and magnetism—the law of electromagnetism:

Magnetic fields are created by moving electrical charges.

Thanks to Oersted, what had formerly seemed to be two independent forces were shown to be intimately linked. But the story of electromagnetism was only half told.

The Law of Electromagnetic Induction: Almost immediately, scientists suspected that a symmetrical version of Oersted's law must be in play. The British physicist and chemist Michael Faraday (1791–1867) announced his breakthrough in 1832 after years of laboratory inquiries. The key experiment, after many failed attempts, involved two coils of wire wrapped side-by-side around an iron rod. Faraday used a battery to pass current through one wire, thus creating an electromagnet. At the same time, the other coil of wire, although not connected to a battery, also developed a short-lived electric current—a phenomenon Faraday called electromagnetic induction. Further experiments firmly established the principle that led to the invention of the electric generator, the electric transformer, and countless other applications that have changed our world. The law, simply stated, is:

Electric fields and electric currents are created by changing magnetic fields.

Together, four statements—the laws of magnetostatics, electrostatics, electromagnetism, and electromagnetic induction—present a unified view of what had once seemed different phenomena. One important challenge remained.

Maxwell's Equations: With the exception of Coulomb's law, the four statements linking electrical and magnetic phenomena are qualitative. They are powerful ideas, to be sure, but not predictive in any quantitative sense.

Scottish physicist James Clerk Maxwell (1831–1879) unified the field of electromagnetism with a set of four equations. Each equation corresponds to one of the four laws. Each is rigorous in its ability to predict the exact magnitudes of the forces and motions caused by intertwined magnetic and electric fields. The mathematics is advanced, and applying Maxwell's equations to real-world problems can be challenging. Yet, taken together, these four statements provide an unprecedented predictive view of the universe—a vision that led to a profound and unexpected insight.

Equations can be manipulated and interrogated in unexpected ways. Maxwell realized that the constant interplay between electric and magnetic fields resulted in a phenomenon not unlike waves radiating outward when a pebble is tossed into a quiet pond. What's more, these electromagnetic waves must travel at 186,000 miles per second—a value dictated by the strength of the electrical and magnetic forces. Maxwell had discovered the very nature of light.

As enlightening as this discovery might have been, a puzzle immediately arose. Every wave has three interconnected properties: its speed, its wavelength, and its frequency. A simple equation relates the three: Speed equals wavelength times frequency. The speed of light is hardwired into Maxwell's equations; it must be 186,000 miles per second. But nothing in the equations appears to restrict the other properties of light waves—their wavelength and their frequency. According to the equations, wavelengths might range from the size of an atom's nucleus to the diameter of Earth. But in Maxwell's day, the only waves known to travel at 186,000 miles per second were visible light, with wavelengths in the tiny range of 150 to 300 billionths of an inch.

Maxwell predicted the existence of an astonishing, unseen realm of invisible light—the electromagnetic spectrum. Discoveries followed quickly. Radio waves and microwaves have wavelengths much longer than visible light. Infrared and ultraviolet radiation bracket visible light, falling just beyond the red and violet wavelengths, respectively. X-rays and gamma rays have much shorter wavelengths than anything we can see. New technologies—radios, microwave ovens, tanning beds, medical diagnoses with X-rays and gamma rays, and countless more—have flooded the marketplace and changed society. And it all started with frogs.

The Laws of Energy

All the laws we've explored thus far deal in some way with forces and motions. The forces of gravity and electromagnetism—remarkably, the only two forces that we experience in any direct way—provide a framework for explaining why objects move. But we're not done. There is at least one other natural phenomenon that plays a vital role in our lives—energy.

Understanding energy was a slow, nonintuitive process, spanning centuries of confusion and false starts. Unlike most of the other universal laws, the principles of energy did not spring forth from the brilliant insights or experiments of a single mind. The reasons probably lie in the abstract, intangible character of "energy"—a phenomenon that eighteenth-century scientists couldn't touch or weigh or measure in any obvious way on a lab bench.

Early in the process of deducing the properties of this strange yet pervasive phenomenon, scientists had to introduce a new technical term—"work." Work is defined by scientists as applying a force over a distance. When you climb a flight of stairs, or stretch a bowstring taut, or hammer a nail, you apply a force over a distance. You do work.

The definition of energy follows: Energy is the ability to do work, that is, the ability to apply a force over a distance. That definition sounds simple, but it is fraught with complexity. Nature holds many different, often subtle ways to do work. No significant progress on the laws of energy could be made before a relatively complete catalog of these many forms of energy had been created.

Kinetic energy, the energy of motion, is the most obvious. Any moving object—think a speeding car or flying cannonball—is able to exert a force over a distance.

Other physical systems have the potential to exert a force over a distance, even if they aren't doing so right now. This potential energy comes in many forms. A boulder perched on the edge of a cliff or water held back by a dam has gravitational potential energy. A stretched rubber band or a tightly wound clock spring has elastic potential energy. A stick of dynamite or the ice cream sundae you ate last night stores chemical potential energy. Magnetic, electrical, and nuclear energy also represent potential energy in their distinctive ways.

That's a pretty long list of kinds of energy. But a major stumbling block to completing our inventory was the mystery of heat. The prevailing view in the 1700s was that heat is a kind of invisible, self-repelling, weightless fluid called caloric that constantly flows from warmer to cooler objects. Somewhat analogous to water, the theory went, caloric soaks into and expands many objects while it spreads out and "evaporates." Every object holds more or less of this fluid: Coal holds a lot, ice almost none at all.

This idea is intriguing, but wrong. The caloric theory's death knell came in the form of a treatise, *An Enquiry Concerning the Source of the Heat which is Excited by Friction*, published in 1798 by the extraordinary inventor Benjamin Thompson (1753–1814), self-styled Count Rumford. Thompson led an adventuresome life. At age nineteen he married a wealthy widow much his senior in Rumford, Massachusetts (hence the adopted title "Count"). He became an active British loyalist and spy during the American Revolution, eventually abandoning his wife and child and fleeing to England. He was knighted by George III, yet subsequently fled England, again on suspicions of spying. He wound up in Hanover in modern-day Germany and was named master of the Bavarian cannon works, where he conducted his most famous experiments.

Thompson helped to discredit the caloric theory by demonstrating that heat and mechanical work are equivalent. Making cannons was key. Each cannon was cast in solid metal and then bored with a long drill. Sharpened drills, Thompson observed, quickly cut through the metal while producing relatively little heat. But try the same operation with a dull drill bit and prodigious amounts of heat could be generated while making little progress in boring. Thompson demonstrated this effect with a public experiment, subjecting an unfinished cannon immersed in water to a dull drill bit, boiling the water in just two-and-a-half hours. If caloric was a fluid contained in the metal, he argued, then rapid boring should release more heat. That didn't happen.

Thompson's conclusion: Heat is a form of motion generated by mechanical work, not a fluidlike substance intrinsic to matter. It took a few decades for this concept to catch on, but eventually it became clear that heat is a form of energy—it confers the ability to do useful work, as epitomized by the steam engine. And so, by 1840 the catalog of forms of energy was filled with kinetic energy, varied flavors of potential energy, and heat. The stage was set for a new round of discoveries.

A vital step in deducing laws of energy was provided by observations on the ways energy can change from one form to another. The gravitational potential energy of a perched boulder can easily be converted to kinetic energy, just as the kinetic energy of an object thrown onto a roof can be converted to gravitational potential energy. Heat can be used to drive chemical reactions (think baking a cake), just as the chemical reactions of a burning candle can release heat. Electricity can be used to generate light, while light can be used to generate electricity. Indeed, virtually any form of energy can be converted to any other using the appropriate technology. But is there something deeper in this confusion of energy transformations?

English physicist James Prescott Joule (1818–1889) is credited with discovering the first law of energy. He designed a clever device that converted precisely measured amounts of energy from one form to another to another. His experiment started with heavy weights—gravitational potential energy. The weights connected to an elegant geared mechanism that turned a paddlewheel, which rotated rapidly in a container of water, thus converting gravitational potential energy into kinetic energy. As the experiment progressed, the paddlewheel's kinetic energy gradually converted into heat, as measured by the water's temperature. At each step of the process, Joule was able to document exactly how much energy was tied up in the forms of gravitational potential, kinetic, or heat. The result: Despite the constant transformations from one form to another, the total energy of the experiment—the sum of gravitational plus kinetic plus heat—was constant.

The first law of energy, or thermodynamics (literally, the movement of heat), follows directly from Joule's experiment:

In a closed system, energy can transform from one form to another many times, but the total amount of energy is conserved.

This first law of energy sounds like pretty good news: It says that energy never goes away. The total is constant.

So, why do we have to pay energy bills?

Second Law of Thermodynamics: We know from everyday experience that nature places some stiff restrictions on how we can use energy. Think of the different kinds of energy as bank accounts. We have the kinetic energy account, the heat account, and maybe an account for gravitational potential energy. We can shift energy from one account to another, for example, by using a clever experiment like Joule's.

What we soon observe is that each transaction—each transfer of energy between accounts—comes with a fee. Some energy is always dispersed—not lost, but unavailable as a bit of heat radiating out into the cosmos. That energy still exists, but it's of no practical use to us.

Examples are all around us, all the time. A hot cup of coffee inexorably cools, making the room just a tiny bit warmer. A significant fraction of our car's full tank of gas winds up as waste heat expelled out the exhaust pipe. We eat a hearty, energy-rich meal, only to find ourselves hungry the next morning.

At the heart of this incessant energy loss is the movement of heat (that is, literally, "thermodynamics"). Heat never sits still. It moves spontaneously from hotter to colder objects by any of three familiar processes.

Heat moves by conduction—the transfer of energy by direct contact between two objects. Accidentally place your hand on a hot stove and you'll feel conduction. (Again, ow!)

Heat moves by convection—the transfer of energy by the movement of gases or liquids. Ocean currents, trade winds, and the cooling fan on your patio all move heat by convection.

Finally, heat moves by radiation—movement from a hotter to cooler object even across the vacuum of space. That's how the Sun keeps Earth warm and habitable.

Humans have found many clever ways to slow down unwanted movements of heat. We insulate our homes, wear down-filled clothing, and design our engines and electronics to minimize loss. But there's no way to stop heat from moving. Hence the second law of thermodynamics:

Heat will not flow spontaneously
from a colder to a warmer body.

This statement is perhaps the most intuitive framing of the second law, but it's not the only one.

The second law of thermodynamics fascinates scientists for

several reasons. For one thing, that simple statement about the movement of heat has profound consequences for our technological world. Indeed, an equivalent framing of the second law is:

You cannot construct an engine that does nothing but convert heat to useful work.

In other words, you can't design an engine that is 100 percent efficient.

One reason for this restriction is simple: Every engine generates heat, and a lot of that heat must escape into the surroundings. If you burn gasoline or coal or oil, the heat generated must go somewhere. But it's more than that. Every engine operates on a cycle—a piston driven by exploding gasoline, for example, where chemical potential energy is converted to heat, expanding gas, and the kinetic energy of the piston, which in turn transfers kinetic energy to the wheels. You can design that piston to be almost frictionless and reduce heat loss in other ways, but you still must expend energy to reset the piston to cycle again. Consequently, a significant fraction of the energy you pay for to operate your car is converted to forms of energy that don't actually move the car. It's important to note, however, that this limitation doesn't apply to every transfer of energy. A space heater, for example, doesn't operate on a cycle. You can convert almost 100 percent of the electrical energy you pay for into heating your home.

Yet another statement of the second law of thermodynamics delves more deeply into the atomic-scale mechanisms of heat:

Every isolated system becomes more disordered with time.

This idea of disorder seems far removed from a cooling cup of coffee or the operation of an engine. Yet disorder is part and parcel of every movement of heat. Think of that hot coffee as a highly ordered collection of energetic molecules that are surrounded by much less energetic (i.e., cooler) molecules. Inevitably, that state of order—an island of hot molecules in a sea of cooler molecules—evens out. The initially ordered arrangement of hotter molecules confined in a cup, surrounded by the colder molecules of the surroundings, becomes disordered as the coffee cools.

We are all too familiar with this natural tendency toward disorder. Your spiffy new leather briefcase will become scuffed, your freshly cleaned living room inexorably gets dusty again, and everyone has had a glass suddenly slip from their hand, falling and shattering into sharp fragments. The more sobering disruptions of hurricanes, tornadoes, wildfires, and floods can cause destruction and death at staggering scales.

All these examples and countless others depend on the concepts of probability and order. To a scientist, order relates to the degree of organization of a system—the extent to which its constituent parts, all the way down to atoms, are arranged. Highly ordered systems, like your new briefcase, tall buildings, a hot cup of coffee, or people, require very strict arrangements of atoms—arrangements that could not possibly arise by chance. Those highly ordered states are improbable, while less ordered atomic arrangements are much more likely. As shoes or buildings or people are subjected to the vagaries of their surrounding environments, their arrangements of atoms must inevitably change. And the probabilities suggest that any change will lead to less order. Scientists quantify this natural trend toward disorder with a measure called entropy. Yet another statement of the second law of thermodynamics is:

The entropy of an isolated system tends to remain constant or increase.

Embedded in these varied yet equivalent statements we find something we haven't seen before in any other natural law. The phenomena described by the second law of thermodynamics, unlike events associated with the laws of motion, gravity, electromagnetism, or even the first law of thermodynamics, have a direction in time. Our lives take place in four dimensions—three spatial dimensions plus time. All the other natural laws are symmetrical with respect to all four dimensions—forward or backward in space or time makes no difference. The elliptical path of a comet, the flow of electrons through a wire, or the transfer of energy from one form to another works equally well both ways.

The second law of thermodynamics is different. It implies that cosmic time has a direction—what physicists call the arrow of time. Heat never flows spontaneously from cold to hot. An engine can't work backward while "unburning" fuel. And your leather briefcase does not spontaneously become unscuffed. We experience the irreversibility of events in profound ways, every day of our lives. We age. We suffer losses. In that careless, distracted instant we slip and fall. A colleague once concluded, "Entropy sucks."

The Second Arrow of Time

Ten laws of nature. Is that all we need? Do they explain everything we experience in the physical universe?

To be sure, there have been extraordinary new advances in modern physics—additional laws that codify physical phenomena at scales we can barely imagine. Physicists have discovered two other forces, called the strong force and the weak force, that only operate in atomic nuclei—phenomena revealed by the radioactivity of some kinds of atoms. The quantum world, in which energy and mass come in little bundles and every aspect of nature occurs in discrete increments, also prevails exclusively at atomic scales. We experience only the vaguest hints of quantum phenomena, for example, in the unnaturally vivid colors of ruby red lasers and Day-Glo paints and dyes.

Relativity is another abstract realm, dealing with incomprehensible paradoxes that occur at speeds close to that of light or masses so enormous that they collapse into a black hole. And we surely have more to learn. Other laws related to the cosmic mysteries of dark matter and dark energy may well await discovery. Nevertheless, we can exclude those amazing discoveries from our list of macroscopic laws—the rules that describe and explain our everyday experiences.

What about more familiar realms? Why, for example, aren't there universal laws of light? After all, the myriad phenomena of light, from the workings of a powerful telescope to the vivid colors of an arching rainbow, seem far removed from any law outlined above. However, in the case of light, Maxwell's four equations codifying electricity and magnetism are all you need. Visible light is simply one flavor of electromagnetic radiation, and the laws of light follow inevitably from the equations.

So, are we done? Can we close the books on discovering macroscopic natural laws? We think not. We think science has missed something obvious—because increasing disorder is not the only aspect of the cosmos that changes with time.

Almost fourteen billion years ago, at the Big Bang's moment of creation, the universe was featureless. There were no stars, no planets, no life. Yet generation after generation of novelty ensued. Atoms and molecules, stars and planets, oceans and atmospheres, rocks and minerals, life and language, and much more have sequentially graced the cosmos. Even as cosmic disorder inevitably has had its way, the universe has evolved and continues to evolve.

Our lives mirror this temporal dichotomy. Yes, we get sick and grow older. All of us must die and our bodies will inexorably fragment, disordering into a vast collection of molecular bits. Yet we also experience countless examples of *increasing* order: stars shine, crystals grow, seeds sprout. Humans create language, music, art, and science. Must not nature's laws echo what all of us experience and know to be true?

We propose that nature reveals an as yet uncodified law of increasing order—a second arrow of time. This missing law must describe and explain the universal tendency for certain systems to display increases in order—to become more interesting, even as the disorder of the surrounding universe increases. Evidence abounds. Look around you: Diverse planets condense from clouds of dust and gas, stunning formations of rock emerge from cooling lavas, abundant flowers bloom in the spring, music and art flow from the human mind.

Generations of previous scientists have suggested that such wonders are simply quirks of a single arrow of time in a universe that is racing to increase entropy as fast as possible. By one extreme version of this disheartening interpretation, our energy-consuming human brains have evolved as an especially efficient way to radiate waste heat into the depths of space.

We disagree. We counter that the breathtaking transformations from atoms to stars to planets to life—what has been called cosmic evolution—are manifestations of a universal imperative that has been at play since the Big Bang. Increased order is fully consistent with all the other natural laws, but it does not automatically *follow* from those laws. What is missing from our canon is a natural law of evolution—a second arrow of time related to increasing order that balances the well-documented first arrow of increasing disorder.

But what form might such a missing natural law take? Can we find a convincing way to describe, explain, and predict the behaviors of nature's wondrous evolving systems?

2

Evolution Everywhere

Evolution. That one simple word has the power to evoke swift reactions and intense emotions. A miracle of biology. A heretical hypothesis. A challenge to teach. A football for theologians to kick around.

The controversies swirling around evolution seem misguided and unnecessary to those of us who study nature for a living. In its most basic guise, "evolution" means change through time—something all of us experience every day of our lives. We contend that evolution is a universal phenomenon, in play since the Big Bang 13.8 billion years ago and continuing today in many systems at many scales. Atoms evolve in stars. Molecules evolve in atmospheres. Minerals evolve on and within planets, as does life in all its varied splendor. What's more, nature's evolving systems often interact with each other: Atmospheres, minerals, and life coevolve in complex, interdependent, and unexpected ways.

But we're getting ahead of ourselves. What exactly is evolution?

What Is Evolution?

Most people, when asked if they "believe" in evolution, immediately think of Charles Darwin (1809–1882)—one of the most influential, and in some circles polarizing, scholars of the last half millennium. In his revolutionary 1859 book, *On the Origin of Species by Means of Natural Selection*, Darwin presented a clear and compelling case for a simple natural mechanism that fosters gradual biological changes. Crafted in stylish prose and filled with accessible observational evidence, *On the Origin of Species* expounded a vivid vision of a biosphere that has been evolving for many hundreds of millions of years.

Darwin rested his theory of natural selection on three well-documented observations about the living world. First, he elaborated on the familiar theme that individuals of any biological species exhibit variations in their traits. Every oak tree has a different array of branches and leaves. Every cherished pet—cat, dog, hamster, or goldfish—has its own distinctive physical features and personality. Each human being is unique.

Darwin's second contention is equally familiar: More offspring are born than can survive. Each individual faces challenges from the moment of birth. Diseases, natural disasters, predators, and more inevitably winnow the herd.

Finally, Darwin argued that the fittest individuals are most likely to survive. Those who are most able to negotiate life's many hazards into adulthood will be preferentially *selected* as the ones most likely to pass their advantageous traits to the next generation.

Taken individually, each of Darwin's three statements is straightforward and unassailable. Individuals in any population differ. Not all offspring survive to adulthood. The fittest individuals are more likely to live long enough to reproduce. There is nothing remotely controversial in any of those observations. But collectively, these three simple ideas imply something that was radical to many of Darwin's contemporaries. Selection for beneficial traits will inexorably, over many generations, turn small changes into entirely new characteristics—the ability to swim, or fly, or walk, or see. What's more, all life on Earth shares a common ancestry dating back billions of years. In his popular science talks, evolutionary biologist Mohamed Noor likes to remind his audience that "no matter what you ate for lunch today—even if it was just a salad—you ate a relative."

Biological evolution is a fact. In layers of rock, we find fossils that record epic changes in our planet's biota—distant ancestors and evolutionary experiments that tell the story of life on Earth. Almost four billion years ago, the earliest life-forms consisted of simple single cells. A half-billion years ago, consortia of cells had begun to cooperate and clump together in multicellular organisms like sponges and jellyfish. Today we have remarkable ecosystems of many interacting organisms—coral reefs, tropical forests, coastal wetlands, and cities. As the quintessential evolving system, life clearly displays those two qualitative traits—increased diversity in a sequence of steps.

Darwin's revolutionary ideas have been universally accepted by the scientific community. To be sure, new layers of understanding have emerged from ever more sophisticated studies in the molecular mechanisms of genetics. We now know that biological information is stored in the form of DNA molecules that are passed from one generation to the next. In the As, Ts, Gs, and Cs (shorthand for the genetic molecules adenine, thymine, guanine, and cytosine) of DNA, we can see small building blocks that have the potential to adopt vast numbers of configurations. That information is shuffled and modified through random mutations and sexual reproduction, which provide ways to produce large numbers of genetic configurations. And those diverse configurations are subjected to selection: Only the fittest individuals survive. Humans have learned to engineer DNA, to probe how genes work, to correct "mistakes" that lead to disease, and to produce new varieties of microbes, plants, and animals—new pathways for evolution of which Darwin never dreamed.

Yet even after more than a century of further discoveries, Darwin's ideas remain foundational to biology. From the rich three-and-a-half-billion-year fossil record of Earth's abundant evolving life, to the ongoing development of new agricultural strains of grains and vegetables, to the unpredictable seasonal mutations of flu and other viruses, Darwin's theory of evolution by natural selection describes and explains how the living world works. In the words of the influential twentieth-century biologist Theodosius Dobzhansky (1900–1975), "Nothing in biology makes sense except in the light of evolution."

We wholeheartedly agree. But whereas many biologists (and other scientists) might also argue the converse—that is, that evolution makes sense only in the context of biology, or that life and its by-products represent the only valid examples of evolving systems—we suggest that the Darwinian model is simply a special case, albeit the most convincingly elaborated one, of a much more pervasive universal phenomenon that has been at play since the Big Bang.

Without doubt, life is somehow fundamentally different from inanimate matter like atoms or minerals. But how we view nature is often a matter of deciding whether to combine different things into one category, versus dividing similar things into separate categories—what Darwin himself called lumping versus splitting in his quest to understand the nature of biological species. Most previous authorities have chosen to split disparate changing systems, placing life in one category and inanimate atoms and minerals in another.

But what if instead we lump? What if we choose to see life and atoms and rocks and planets and atmospheres and language and science as one connected, universal phenomenon of cosmic evolution? Consider the possibility that both living and nonliving systems become more complex and ordered through time. Armed with that perspective, we might find ourselves on a path to glimpse the elusive missing law.

In that spirit, let's step back from the particulars of biology and go in search of a more universal definition of evolution and evolving systems. We might see evolution as a process by which a system becomes more diverse, more complex, and more interesting with time, with that increased complexity occurring in a sequence of logical steps—each one building on what came before.

The path forward depends on looking in detail at the behaviors of varied natural evolving systems. There's no better place to start than with the most ancient big evolving systems of all—the first stars to form after the Big Bang.

The Evolution of Atoms

Cosmic evolution began at an instant of time—a moment of creation 13.8 billion years ago when everything that defines our universe appeared, seemingly out of nothing. All matter and energy simply became. The nascent cosmos was compressed into a volume so small that all forces, mass, and energy were homogenized into an incomprehensibly hot, dense dot. For now, that scientifically unsatisfying description is about as good as cosmologists can do.

But scientists have been able to provide a remarkably detailed description of what happened next in a vivid narrative of a rapidly changing universe—dramatic events in the seconds and minutes following the Big Bang. All known forces—gravity, the electromagnetic force, and the strong and weak atomic-scale forces—emerged in the universe's first fraction of a second. The fundamental building blocks of atoms—protons, neutrons, and electrons—all followed soon thereafter, while the first atomic nuclei with simple combinations of a few protons and neutrons appeared when the universe was about three minutes old. No stars or planets arose at that early stage, and nothing resembling what we think of as the universe evolved for a very long time. Everything existed as a soup so scalding that positively charged nuclei and negatively charged electrons couldn't bind to form atoms. In such a milieu, photons—particles of light—ricocheted feverishly off freely swimming electrons. It was far too hot, far too chaotic, for order to arise.

The next half-million years saw the cosmos rapidly expand and cool. Once protons and electrons slowed down enough for atoms to form, the photons of light could finally wander freely. We see the vestiges of this primordial light as a faint glow of microwave radiation that permeates the universe today, telling the story of the moment when atoms, almost all of them hydrogen or helium (the first two elements of the periodic table), emerged some four hundred thousand years after the Big Bang. Only then could gravity begin to lump atoms together to make the first generation of stars—the engines of atomic evolution.

Atomic Anatomy: To understand the evolution of atoms, it's helpful to review how they are put together. Three familiar building blocks play their distinctive roles.

First comes the proton—a relatively massive particle in the

nucleus that sports a positive electrical charge. Protons are important because their numbers define each chemical element. Hydrogen is element number one, with exactly one proton. Helium is the second element, always with two protons. Carbon invariably has six protons, iron twenty-six, and gold seventy-nine.

The other familiar nuclear particle is the electrically neutral neutron, just a tad more massive than the positive proton. The number of neutrons in an atom doesn't change the name of the chemical element. An atom with two protons and one neutron is helium. So is an atom with two protons and two neutrons. But, in a useful bit of atomic bookkeeping, we catalog these two varieties of helium as different isotopes: two protons plus one neutron is called helium-3, while two protons plus two neutrons is helium-4. If you know the element name and the isotope number, you pretty much know what you need to know to tell one atom from another.

The third essential atomic building block is the fickle, negatively charged electron, which leads its frenetic life far from the nucleus. Electrons surround nuclei in different shells, or energy levels, though electrons in the outer fringes of an atom commonly shift their addresses from one atomic home to another.

In thinking about the evolution of atoms, our principal focus is the nucleus and all the many possible combinations of protons and neutrons. Nature reveals that a few combinations are allowed, but the vast majority are not. Remarkably, the great mixing bowls of nuclear particles, where new recipes are tested, are the stars.

Stars—The First Generation: Appearing a half-million years after the Big Bang, the first stars were seemingly simple. These giant gravitationally bound spheres held about nine hydrogen atoms for every helium atom, with only a smidgen of a few other elements, all held together by the force of gravity. Hydrogen came in two flavors, with more than 99.9 percent hydrogen-1 (that is, nothing more than isolated protons without a neutron) and the rest hydrogen-2, a coupling of a proton and a neutron with its own special name, deuterium. A pair of isotopes also represented helium, mostly helium-4 with a smattering of helium-3.

Then the fun began. Stars are engines of atomic evolution because their internal temperatures and pressures are incomprehensibly high—so great that isotopes smush together, colliding to combine in nuclear fusion reactions. In every star, the first stage of this remarkable atom-generating process (astrophysicists call it stellar nucleosynthesis) is hydrogen burning. Hydrogen atoms collide again and again, fusing to generate helium-4—a process that not only creates new atoms but also releases prodigious amounts of energy. Next time you lie on a sunny beach, soaking up rays, thank hydrogen burning in our nearby Sun.

Any star as large or larger than the Sun gradually uses up its supply of hydrogen while generating more and more helium-4. That buildup of helium-4 eventually leads to helium burning, the next in a sequence of atom-making recipes. Triplets of helium-4 combine into carbon-12—a process that created the universe's first significant pulse of the element of life, element six. A cascade of fusion reactions follows: Carbon burning reactions employ carbon-12 to form oxygen, magnesium, neon, and other elements. Oxygen burning generates silicon, phosphorus, and sulfur from oxygen-16. A stepwise sequence of fusion reactions combining those and other elements with helium-4 leads to a cascade of new elements, each with two more protons and two more neutrons than the last.

Fusion reactions work because each newly generated isotope holds a bit less energy than the sum of its smaller building blocks. As codified in the first law of thermodynamics, that extra energy must be conserved—it must wind up somewhere, and so it becomes heat. That heat, coupled with a star's immense internal pressure, drives even more fusion reactions in a self-sustaining process that can persist for billions of years.

All in all, the fusion reactions in the first generation of stars after the Big Bang produced about a dozen different chemical elements in abundance, with smaller amounts of maybe a score more. But if that were the whole story of stellar nucleosynthesis, then you wouldn't be reading this book. Almost all the atoms that make our lives interesting would still be locked inside those first-generation stars. To be sure, a modest fraction of a star's mass gets blown out into space in the form of stellar winds. Small quantities of carbon and oxygen and other raw materials of life began to accumulate in the depths of space, but not enough to make planets and life. One of the most dramatic processes in the cosmos had to come next: Stars had to explode.

Exploding Stars: The life story of a star depends on its mass. Modest-sized stars like the Sun saunter through three prolonged stages, at first burning hydrogen to make helium—a reliable process that can last billions of years. That's a good thing for those of us living on planet Earth. Our Sun has been stable for more than four billion years, long enough for life to evolve in wondrous ways. Astrophysicists tell us that we can expect at least another four or five billion years of solar stability, albeit with the Sun ever so gradually heating up such that Earth's surface may become uninhabitable in another one to two billion years.

Inevitably, the Sun's supply of burnable hydrogen must dwindle to the point that helium burning, a faster and more intense process, takes over. Our once-stable Sun will heat up much more rapidly than before, ballooning outward to many times its current size to become a red giant star. The innermost planet Mercury will be swallowed whole, and maybe Venus as well. Earth, its sky overwhelmed by the immense, looming red star, will be scorched and blasted beyond recognition. For a half-billion years, a fat ruddy Sun will consume its stores of helium, producing an ever-greater stockpile of carbon.

And then, for our Sun, it will be game over. Consumable helium will be spent. The Sun is not quite massive enough to sustain the fusion of carbon atoms. Lacking the immense outward pressure generated by nuclear fusion reactions, the Sun will collapse, shrinking smaller and smaller to a planet-sized, carbon-rich sphere. The Sun will become a white dwarf star.

Earth will still orbit that massive little object—a dense carbon-rich star that glows blue-white hot yet produces less than a thousandth of the radiant energy we enjoy today. Our planetary home will have become a baked, desiccated world.

Don't worry too much. That's still a few billion years down the road.

▪

Stars larger than the Sun experience a different fate. Their greater internal gravitational forces generate internal pressures and temperatures sufficient for carbon-12 to fuse into a heavier suite of isotopes—oxygen-16, neon-20, magnesium-24, and more. The ensuing cascade of nuclear reactions endows the star with an ever-expanding inventory of chemical elements. Each new stage is more urgent; each results in more insistent outward pressures pitted against the force of gravity. A frenetic sequence of fusion reactions defines the final stages of all large stars, taking only seconds to change the universe forever. In a moment, the profusion of fusion reactions all but stops. Production of iron-56 is the final step.

Each element less massive than iron generates energy as it forms. Each of those elements is fuel that allows the star to shine. Iron-56 is different. Like the ash in your cold, dark fireplace, iron-56 has no more energy to give. There's nothing you can do to tease out an extra bit. Add or subtract a proton or a neutron from iron-56 and the resulting isotope always has more energy, not less. Like a dying fire bereft of fuel, the star turns off. The fusion reactions countering gravity cease. Gravity takes over with a vengeance.

A large star's catastrophic collapse is beyond imagining. All its mass, from the unburnt storehouse of hydrogen and helium to the freshly forged inventory of iron, plunges inward and smashes together, the collapsing mass approaching the speed of light. The resulting pressures, temperatures, and densities exceed anything experienced since the dawn of the cosmos—since the Big Bang itself. Isotopes of every kind mix and fuse to produce previously unknown arrangements of protons and neutrons, with some heavy isotopes boasting more than two hundred protons and neutrons. And so it was, more than thirteen billion years ago, when the first stars exploded, and the cosmic catalog of isotopes surged.

Big stars not only collapse; they tear themselves apart, exploding in a supernova. In the process, the newly expanded atomic inventory is flung far out into space. Not until the first generation of big, iron-rich stars exploded could rocky planets form. Not until those stars flooded the universe with carbon could life emerge.

Stars—The Next Generation: The logical sequence of fusion reactions in stars—hydrogen burning, helium burning, carbon burning, and so on—can only get you so far in the evolution of atoms. Two other processes, both involving vast numbers of neutrons, help to round out the isotopic zoo.

The second generation of stars, as well as all the generations that have followed, began their lives with lots of iron—an element forged abundantly in the first big stars and scattered about the cosmos via supernovas. Iron promotes element diversification by soaking up neutrons, in what is known as slow neutron capture, or the S process.

Many elements have the potential to accommodate extra neutrons, which are constantly produced during fusion reactions. When a neutron collides with an existing nucleus, it often just sticks there, increasing the isotope number by one. Many atoms aren't disrupted by the addition of a few neutrons. Iron-56 plus one neutron leads to stable iron-57. Add another neutron and you have stable iron-58. But many other iron isotopes, all the way from iron-59 to iron-72 (with a whopping forty-six neutrons), are not stable. Each of them has been made in laboratories, and each is surely generated in stars as well. Such excesses of neutrons result in unstable iron atoms that undergo radioactive decay into myriad new isotopes, some of them stable forms of entirely new elements.

A second neutron capture process, the rather mysterious rapid or R process, generates even more isotopic diversity in stars. Fully half of the isotopes heavier than iron are thought to arise only in this way. Yet the cosmic mechanisms of rapid neutron capture events have proven difficult to ascertain, largely because they require an astonishingly large neutron flux—a concentration of neutrons not found in any ordinary star. Hypotheses range from formation in the cores of the largest supernovas to so-called kilonova collisions, involving neutron stars. Whatever the mechanisms of R-process synthesis, the final violent stages of stars' lives generate the full spectrum of natural elements and isotopes.

■

The evolution of stars leads inexorably to the evolution of atoms—processes that epitomize nature's tendency to become more diverse and more interesting. Starting with a handful of small isotopes, stellar nucleosynthesis produces one generation after the next of atomic novelty, each generation displaying greater isotopic diversity than what came before. The cosmos also enjoyed the fruits of a sequence of element-forming processes. Life-giving carbon could not have formed until the prolific production of helium. Silicon and oxygen, critical to mineral evolution, had to wait for the synthesis of carbon, while iron came later still in that logical process. And that was only the first generation of stars.

Entirely new stellar populations, rich in elements heavier than hydrogen and helium, continued the sequence of atomic evolution. The S process and R process worked their magic on iron, then on the myriad by-products of iron, each new isotope appearing in its turn. From two primordial Big Bang elements, hundreds of isotopes eventually emerged.

The evolution of atoms led to what might seem to be an inexhaustibly rich diversity of atomic building blocks. Yet, from another point of view—the perspective of combinatorics—the inventory of atoms and isotopes is exceptionally parsimonious.

Combinatorics: A remarkable feature of atomic evolution—a feature shared by all evolving systems—is the paucity of different realized kinds. We propose that all evolving systems share three characteristics—(1) the potential for vast numbers of configurations, (2) processes to generate numerous configurations, and (3) a selection process that rewards the tiny fraction of all possible configurations that actually "work." In the words of Saint Matthew, "Many are called but few are chosen."

Isotopes are a case in point. Nature provides us with only about 340 different stable isotopes, distributed among only 90 stable chemical elements. The overwhelming majority of arrangements of protons and neutrons decay away or never occur at all. But how many combinations might there be?

The heaviest naturally occurring isotope is uranium-238, with 92 protons and 146 neutrons. It's easy to calculate all the possible combinations of protons and neutrons from only 1 to 238. Think about the sequence. With only one particle there are two possibilities—a proton (i.e., hydrogen-1) or a neutron (which is unstable and doesn't hang around very long by itself). With two particles, the number of combinations increases to three: 2 protons, or 2 neutrons, or 1 of each. Only the latter is a stable isotope, hydrogen-2 or deuterium. Three particles lead to four possibilities, four particles lead to five, and so on up to 238 particles with 239 different possible combinations of protons and neutrons.

To get the grand total, just add all the numbers: 2 + 3 + 4 + . . ., all the way to 239. And the answer is . . . wait for it . . . 28,649. It turns out that only about 1.2 percent of all possible combinations of protons and neutrons—340 out of 28,649—occur in nature. Stars, the giant mixing bowls of nuclear particles, undoubtedly sample all possibilities at one time or another, but only a small fraction persists. Thus, the evolution of atoms in stars displays all three important characteristics:

1. Atoms form from vast numbers of protons and neutrons, which have the potential to adopt large numbers of different configurations.

2. Stars are immense mixing bowls, providing a natural mechanism by which many of those different configurations are generated.

3. The requirement for stable atomic nuclei results in the selection of only a small fraction of all possible configurations.

The Evolution of Minerals

The evolutionary imperatives that generated atoms long ago were the first steps in an ever-expanding saga of cosmic chemical evolution. As the universe expanded and cooled, and as stars flooded space with novel bits of matter, new arrangements of atoms, called planets, graced creation.

Before the first generation of big stars exploded, collections of atoms invariably took the form of a gas or a plasma. A gas, like the air we breathe, is a localized collection of atoms and small molecules that are too warm to settle down. The tiny gas particles fly around with a lot of kinetic energy, bumping into each other, never stopping to rest.

A plasma, the stuff of stars, is a lot like a gas, but much hotter, with more frenzied collisions—impacts so violent that some electrons are stripped off the atoms and join in the melee. The strange result is a gas-like state with lots of positively charged atoms in a turbulent sea of negatively charged electrons. Those mobile electrons conduct electricity, while the plasma can be pushed and prodded by a well-shaped magnetic field.

Early in the history of the universe, matter was too hot, and molecules too small, for any organization of atoms other than gas or plasma. Nevertheless, another state of matter was waiting in the wings for its turn to evolve. Minerals and rocks, the solid foundations of terrestrial planets and keystones for the origins of life, were soon to follow.

Cosmic Crystals: The oldest minerals in the cosmos, the first solid matter of any kind, formed in the expanding, cooling atmospheres of the most ancient generation of stars—the first stars following the Big Bang. All it took was for hot stellar atmospheres enriched in a few common elements—carbon, oxygen, magnesium, and silicon chief among them—to cool below a few thousand degrees. In time, a suite of sturdy, high-temperature minerals condensed and grew as microscopic crystals.

According to some models, microscopic diamond crystals—pure carbon condensed into stardust—formed the first mineral species in the cosmos. (In mineralogical parlance, a species is a natural crystal with a unique combination of chemical composition and crystal structure; diamond, for example, is the element carbon arranged in the dense diamond crystal structure. Today, the International Mineralogical Association has approved more than 6,000 distinct mineral species.) A few other microscopic stardust minerals followed, including olivine, corundum, moissanite, and spinel—all valued as gemstones in their much larger terrestrial forms. Eventually, about 25 different stellar mineral species became the critical building blocks of countless subsequent generations of planets, including Earth and all other rocky worlds.

Our solar system was born 4.567 billion years ago when a great mass of hydrogen and helium gas, well mixed with mineral-rich stardust, began to collapse under the force of gravity. Most of that mass—99.8 percent by reliable estimates—

wound up in our central star, the Sun. Most of the rest became the planets, but not without a prelude of dramatic mineral-forming events.

The earliest Stage One of our solar system's mineral evolution began as electrostatic forces caused microscopic bits of stardust to coalesce (think dust bunnies in space). As the Sun's internal nuclear fusion reactions began to release energy in earnest, intense bursts of heat caused those fluffy clumps to melt into little droplets, called chondrules. As the droplets subsequently cooled and froze, a new generation of melt-grown minerals appeared—as many as 100 different kinds of crystals.

The inner solar system generated vast numbers of chondrules, while gravity held sway and more clumpiness ensued. Irregular rounded rocky objects the size of basketballs, and then football fields, cities, and states, grew. Epic collisions between planetesimals generated larger and larger bodies as bigger masses inevitably swallowed smaller ones. The resulting violent second stage of mineral evolution saw the large-scale alteration of Stage One minerals, as new kinds of crystals arose from extreme heating, through interactions with liquid water, and from the intense shock of impacts. Among the Stage Two mineralogical novelties were the first clay minerals (the raw material of pottery), the first calcite (the white mineral of marble), and the first crystals of halite (what we sprinkle on food as table salt). Within its first million years, before the first planets appeared, our solar system's mineral inventory had jumped to more than 300 species.

Earliest Earth: The frenetic pace of planetesimal collisions could not last for long. Eventually, a few of the largest objects won, having consumed all but scattered rocky remnants (some of which fall to Earth today as meteorites). The result of all this consolidation? Mercury, Venus, Earth, and Mars—the four mineral-rich planets closest to the Sun.

Planet formation led to new stages of mineral evolution. Earth's first homegrown minerals formed as molten lava poured from volcanic vents, cooling, crystallizing, and covering our planet's surface in thick layers of black basalt, much like the landscapes of Hawaii and Iceland. Basaltic rocks hold what are still some of Earth's most common rock-forming minerals—feldspar, olivine, and pyroxene. These vents also released water vapor, carbon dioxide, and other gases into the rapidly forming atmosphere and nascent oceans—fluids that altered the first generation of minerals into more than 450 species. Earth's first pliable mica group minerals, the first layers of carbonate rock, and the first ice (yes, ice is a mineral species—H_2O in the ice crystal structure) must have appeared within the first million years.

A little later in Earth's turbulent history, perhaps about fifty million years after its formation, our planetary home suffered its greatest indignity. A Mars-sized object, much smaller than Earth but more than large enough to wreak havoc, impacted the young planet. Earth's surface was obliterated. All near-surface rocks were vaporized and blasted into space. Most of that incandescent mass rained back to form a globe-spanning magma ocean, while a smaller fraction consolidated to form the Moon. Like a giant reset button, every mineral on Earth was destroyed in a kind of mineralogical mass extinction.

Quickly, perhaps in just a few thousand years or less, the minerals started to return, one by one, in a replay of Earth's earliest mineral evolution. At first the magma cooled to igneous rocks with their 100 or so mineral species. Then a second generation of oceans and atmosphere altered those minerals, giving birth to the full complement of familiar species that populated Earth before the Moon-forming cataclysm.

Once it had settled back down, Earth was poised for a dramatic expansion of its mineralogical repertoire.

Paragenetic Modes: Through more than four billion years of change, Earth has devised one new way of forming minerals after another. Each mineral-forming mechanism—each "paragenetic mode"—has led to new species.

Melting and freezing played major roles. Rocks, which are intimate interlocking mixtures of as many as a dozen different minerals, have a peculiar property. When heated, they don't melt all at once. Rather, one or two minerals melt first, thus creating a magma that is quite different in composition from the bulk rock. A case in point is that durable black rock basalt, which solidified as lava poured from Earth's earliest volcanoes and blanketed the surface. Basalt, when buried, heated, and partially remelted, produces an entirely different kind of lava—one that is much richer in silicon and aluminum, poorer in magnesium and iron, and much less dense than basalt. The resulting magma rises buoyantly toward the surface to make a new kind of igneous rock called granite.

The magma that produces granite also concentrates a host of rare elements—valuable resources like lithium, beryllium, boron, uranium, and the so-called rare earth elements that are essential to modern information and energy technologies. Where rare elements go, new minerals are sure to follow. Earth's early granites may have hosted as many as one thousand mineral species, many of them new minerals that could not have formed before basalt had partially melted.

Another two-thousand-plus new minerals likely arose in conjunction with the emerging global process of plate tectonics. Earth, unlike any other planet in our solar system, displays dynamic interactions between its rocky skin—the outer few tens of miles that includes Earth's crust—and the underlying hot, soft, deformable upper mantle, which extends down about 250 miles. Relatively thin, brittle sheets of rigid rock called plates slowly move across Earth's surface, shunted about because of the grand cycles of deep mantle convection (think of a boiling pot of water in very slow motion).

As tectonic plates collide, they interact in dramatic (i.e., mineral-forming) ways. Some subducting plates plunge down into the mantle, where they partially melt, producing new kinds of volcanism with associated metal-rich, ore-forming fluids. Hundreds of new minerals of valuable metal elements—copper, nickel, silver, gold, and more—have been discovered in the associated economically valuable deposits. Colliding plates also form new mountain chains, as masses of rocks from deep crustal reservoirs are uplifted, crumpled, and contorted. These metamorphic rock formations feature as many as a thousand new mineral species, including colorful varieties of garnet, lustrous amphiboles, and glittery micas that represent high-pressure and high-temperature alteration products of prior generations of minerals.

All in all, Earth manufactured more than four thousand mineral species through a succession of physical and chemical processes—evaporation, freezing, melting, weathering, and more—each process reworking and refining what had come before. Yet recall that Earth today boasts more than six thousand mineral species. What paragenetic modes have we missed? The answer is linked to another of Earth's chemical wonders—life.

The Coevolution of Minerals and Life: Of Earth's myriad mineral-forming tricks, abundant, diverse, and evolving life sets our planet apart from all other known worlds. Life doesn't simply evolve by itself; it coevolves, both exploiting and changing the atmosphere, the oceans, and the minerals, too.

When lava erupts, chemicals from Earth's deep interior flow over the land. Not surprisingly, Earth's interior is not in chemical balance with the surface. Deep rocks are richer in electrons than surface rocks. So, like the two ends of a flashlight battery, electrons tend to flow from newly erupted basalt into its surroundings in a gentle electric current. All a microbe has to do to get a free lunch is sit on an electron-rich basalt mineral (iron-bearing olivine and magnetite are common examples) and act like a circuit to the environment—think of each cell as a little lightbulb. As the minerals' excess electrons flow into the oceans and atmosphere, the minerals alter to new species that are more at ease in the surface world.

The emergence of life profoundly affected Earth's mineral diversity, but not for a long, long time. For perhaps a billion years, the first generations of primitive microbes sped up chemical reactions that would have occurred anyway, albeit much more slowly. These single-celled organisms altered black iron-bearing minerals to rusty-red iron oxides, they precipitated new layers of carbonate-rich limestone, and they concentrated phosphorus into a variety of phosphate minerals—all chemical reactions that would probably have occurred anyway. For hundreds of millions of years, thriving microbial colonies persisted at or near Earth's surface, making a living on the rocks.

A recurrent theme in all evolving systems is the tendency to experiment, to try new configurations, and to find better and more reliable ways to persist. No energy resource in the solar system has been more robust and reliable than our nearby star, the Sun. It took hundreds of millions of years, but life eventually learned to exploit that radiant energy source. In a sequence of remarkable evolutionary advances, cells developed a complex cascade of chemical reactions that use sunlight to convert water into hydrogen fuel and the reactive waste gas oxygen.

Oxygen-producing photosynthesis started slowly, probably prior to 3 billion years ago, in places where shallow water and abundant nutrients fostered growth. The earliest whiffs of oxygen seem to have been localized, transient, almost like tentative experiments in a new way of living. But by 2.5 billion years ago, the start of an extended time called the Great Oxidation Event, the process had gone into high gear, flooding the atmosphere with oxygen.

The mineralogical consequences of the Great Oxidation Event were immediate and profound. Like an iron utensil left outside, the planet rusted. Earth's earliest oceans were rich in dissolved ferrous iron—a chemical state of iron that holds one restless electron just waiting to be given away. For a billion years, the oceans were perfectly happy to host immense quantities of dissolved ferrous iron.

Oxygen, by contrast, desperately seeks electrons. That's why metals rust and fires burn. Inevitably, as the oxygen content of the atmosphere rose, the ferrous iron in the oceans reacted. The resulting by-product, electron-poor ferric iron, is not soluble in water. This electron-saturated version of iron readily bonds to oxygen, forming a suite of dense iron minerals that sink to the ocean floor. In this regard, iron is not alone. Dozens of chemical elements other than iron play a similar game. Before the Great Oxidation Event, the normal state of many elements was to be electron-rich, poised to give them away. Hundreds of minerals containing copper, cobalt, manganese, nickel, uranium, sulfur, arsenic—indeed, more than a third of the elements of the periodic table—began to alter, to oxidize in the evolving atmosphere. What followed was the greatest expansion of mineral diversity in Earth history—more than two thousand new minerals, a third of all known species, appeared over the next few hundred million years. All those new minerals were the consequence of oxygen released by photosynthetic cells.

Life has influenced minerals in other remarkable ways. For almost all of Earth history, the lands have been lifeless and barren. For Earth's first four billion years, life thrived exclusively in the seas. Why venture onto land when you're doing fine underwater?

Another rule of evolution is that complex evolving systems discover novelty—they try myriad new configurations. In the process, they learn to persist in new ways and exploit new opportunities. Nonlife became life, perhaps the ultimate novelty. Life learned to gather the Sun's energy. Cells learned to cooperate in communities and multicellular structures. In that evolutionary context, it's not surprising that life ventured onto land roughly five hundred million years ago.

Several of life's innovations had mineralogical consequences. For one thing, life learned to use minerals to make a home. More than half a billion years ago, worm-like organisms constructed crude mineral-lined protective burrows. Soon thereafter, symmetrically sculpted mineralized shells appeared, and then teeth and bones. In a kind of biological arms races, bigger mineralized teeth and tougher mineralized shells evolved in parallel—all innovations initiated in the oceans.

Life on land began with tiny, rootless plants that had to grow close to sources of water. That limitation eased with the invention of roots about four hundred million years ago. The first roots were tiny things, a fraction of an inch long, supporting a plant not much taller. But evolution moved quickly. Root systems became ever better at dissolving rocks and creating soil. Deeper roots allowed taller plants, and thus an advantage in the insatiable quest for more sunlight. By three hundred million years ago, forests of towering tree-like ferns covered the land. Again, the mineralogical consequences were dramatic. Roots break apart rocks, in the process manufacturing immense quantities of clay minerals. A new suite of biologically generated clay-like minerals emerged, forming deep soils and transforming the landscape.

Life also played direct roles in generating mineral diversity. Tree-like plants of the earliest forests died, accumulating in thick carbon-rich deposits. Some new minerals arose from plant decay. Others appeared when plant remains were buried, heated, and squeezed to become layers of coal, with new kinds of carbon-rich organic minerals. Even more mineral diversity emerged when a coal deposit caught fire, perhaps ignited by lightning strikes. The resulting hot, burning coal, like the refiner's fire of a blast furnace, "smelted" surrounding rock and produced hundreds of previously unknown crystalline phases.

Animals did their bit, as well. Urine and bird droppings, rich in the mineral-forming chemical ammonia, led to dozens of new minerals. A favorite is spheniscidite, a rare mineral species only known to form on Elephant Island in the British Antarctic, where penguin urine seeps into the underlying clay-rich soil. The mineral name is from *Sphenisciformes*, the technical term for the order of penguins.

Penguins are by no means the end of the mineral evolution story. At a time when human impacts on Earth's environment and climate are under ever-increasing scrutiny, anthropogenic minerals are seen by some scholars as a new stage in Earth's mineral evolution. For thousands of years, humans have learned to make building materials that mimic rocks: bricks, cement blocks, asphalt, concrete, and other sturdy, interlocking mixtures of mineral-like grains. High-tech applications have led to hundreds of new crystalline objects: semiconductor microchips, laser crystals, synthetic gemstones, pressure sensors, magnets, phosphors, abrasives, and more. So ubiquitous, distinctive, and long-lasting are these products of technology that some geologists want to use them to define a new geologic age—the Anthropocene Epoch. Long after we are gone, they argue, the detritus of our society will form an unambiguous marker layer of exotic sediments—the buried remains of cities and towns across the globe.

Through every stage of Earth's storied past, minerals display all the hallmarks of an evolving system. Minerals diversify in a series of logical sequential steps because of three now familiar traits: (1) they are composed of scores of different mineral-forming elements with the potential to adopt vast numbers of different configurations, (2) Earth has many ways to mix and match those elements to sample lots of new configurations, and (3) a tiny fraction of all possible configurations results in a stable crystalline form—a new mineral.

Evolution as a Lawful Process

Evolution is a theme with many variations. For billions of years, atoms and isotopes have evolved in generation after generation of stars; the richness of the periodic table emerged from the simplest beginnings of hydrogen and helium. The rise of the mineral kingdom through billions of years of Earth history epitomizes evolution, with new crystalline species appearing in varied environments through the actions of scores of processes, on land, in the oceans, and deep beneath planetary surfaces. Life evolves, as well, in the most flamboyant explosion of novelties known in the cosmos.

Our bold contention is that all these physical and chemical systems, and many more as well, are examples of the same lawful process—a universal imperative for natural systems to increase in diversity while following a logical sequence of congruent steps. At first blush, these natural evolving systems seem quite different from one another. Atoms and isotopes emerge from the extreme conditions of stellar interiors, where protons and neutrons shift and shuffle in energetic nuclear fusion reactions. Mineral evolution relies on an even greater range of reactions among scores of chemical elements in the diverse temperature and pressure environments of planets and moons. And life, in all its extravagant variety, evolves by its own distinctive set of Darwinian rules. How can there be any underlying patterns in such demonstrably different systems?

We have seen that these examples (and, we would suggest, many, many others) are conceptually equivalent in three important respects:

1. First, each of these systems is formed from many interacting pieces. Those pieces differ significantly in their sizes and behaviors: nuclear particles, chemical elements, organic molecules, biological cells, words, or individual members of a society. Nevertheless, in every instance the individual pieces can be shuffled into huge numbers of different arrangements. Every evolving system shares this common trait: *The potential exists for lots of different configurations.*

2. Second, each of these systems is dynamic, with processes that generate lots of different configurations. Protons and neutrons are incessantly shuffled by stars. Elements are constantly rearranged by planets. Life also plays the game, with new configurations of cells (and their genes) produced and tested in the struggle for survival. Evolution only progresses when *new configurations are actively being generated.*

3. And third, some configurations, by virtue of their stability or dominance or other "competitive" advantage—what we call their function—are more likely to persist. Only isotopes that don't decay, or minerals that don't disintegrate, or individuals who don't die before reproducing themselves, survive in ways that we can study and catalog them. *Selection winnows out most configurations.*

These three characteristics suggest an important conceptual equivalence in the way evolving systems are structured and how they behave. Each system evolves via the selection of a few advantageous configurations out of many possibilities—systems for which most arrangements are duds.

Further, we suggest that such a universal evolutionary imperative is not yet codified in any of the accepted macroscopic laws of physics. This missing law of nature, if it is in some ways true and awaiting discovery, must be in every way consistent with the other laws—the laws of motion, gravity, electromagnetism, and energy. But this hypothetical missing law doesn't follow logically, inevitably from any combination of those other profound statements. It is something different, something new.

Nevertheless, it is not immediately obvious if, much less how, the three characteristics of evolving systems we have observed can be translated into such a law of nature. To understand that connection, we need to delve more deeply into what exactly we mean by "selection" and "function."

3

Selection and Function

We live in a universe that constantly generates marvels—systems featuring captivating structures with elegant patterns and complex behaviors that change over time. We posit that the current catalog of natural laws does not adequately describe or explain the origins and evolution of such systems. We further suggest that any such law—any universal description and explanation of evolution—must by its very nature embed a direction in time. Scientists call such directionality an "arrow of time." The universe began with hydrogen and helium; now we have almost one hundred chemical elements. Stars initially produced a couple of dozen minerals; now we have more than six thousand. Life began as primitive single cells; now we have redwoods and whales, the Amazon jungle and the Great Barrier Reef. In an evolving universe, the present is inherently more complex and entertaining than the distant past.

Modern physics typically explains the directionality of time by invoking what is known as the past hypothesis, which suggests that the universe began in an extremely dense, concentrated, and uniform state (what we would call a low-entropy state). In other words, immediately after the Big Bang the universe was as homogeneous and orderly as possible. The past hypothesis, in combination with the other laws of nature, leads directly to the second law of thermodynamics: The disorder (i.e., entropy) of the universe tends to increase from that initial state.

Let's do a thought experiment. Imagine a possible universe with the same initial uniform, low-entropy state as our own—a universe that unfolds following its own version of a Big Bang. Imagine that universe marching through time in full accordance with the second law of thermodynamics. But this imagined universe is different. It has no attractive force of gravity to pull matter into stars and planets. This universe has no electromagnetic forces to create chemical bonds—no incentives to forge molecules or minerals or life. This fictional universe has absolutely no attractive forces of any kind. In such a universe evolution cannot play a role.

On the contrary, in this patternless universe of our imagination systems rapidly march toward states of higher disorder without generating any long-lived pockets of order. No atoms, nor molecules, nor minerals can arise, much less the complexities of stars, planets, or biospheres. Absolutely no barriers exist that might prevent that boring universe from taking a direct path to a dispersed, disordered state of maximum entropy. Indeed, in such a depressing universe no boundaries can be drawn between different entities—different kinds of rocks and minerals, for example. In such a bland, non-evolving universe, nothing can be distinguished, and nothing can emerge as different. This entire imaginary universe is just a featureless soup of matter and energy, quickly cooling off and dispersing into the void.

Our universe is emphatically not that universe. (Whew!) Our constantly evolving cosmos produces endless entities that do not take the most direct paths to their disordered, high-entropy states. Instead, nuclear forces lead to atoms. Gravity leads to stars and planets. Electromagnetic forces lead to the chemical bonds of minerals and molecules and life. Meanwhile, something intrinsic to our universe—something fundamental about our natural world—"frustrates" the rapid dissipation of energy and the equally rapid increase in entropy.

Many such barriers to achieving maximum entropy exist. For nuclear fusion, limitations in the ways new isotopes can be manufactured in stars, coupled with the inherent stability of atoms like hydrogen and helium, prevent all the hydrogen in the universe from spontaneously alchemizing into the most stable of all the isotopes—iron-56. For chemical reactions, similar barriers of slow reaction rates, coupled with the inherent stability of atoms and small molecules, keep chemical systems from cascading to their lowest energy states. The second law of thermodynamics tells us that the prodigious stored heat energy of hot, molten planetary interiors must move outward to the cooler surface and thence to the vast coldness of space. But that transfer of energy takes many billions of years, thanks to another barrier—the slow movement of heat through dense rock.

On the Nature of Selection

The astonishing diversity of materials and objects in our universe is mute testimony to the effectiveness of nature's barriers to disorder. The elements of the periodic table exist because light isotopes do not easily fuse together, nor do heavy isotopes split apart, to form iron. Minerals forged at the pressure and temperature conditions of Earth's mantle can persist on the surface because they are metastable—resistant to alteration. Life presents even more dramatic examples of persistence in the face of increasing disorder. The vast majority of organic molecules in your body are not very stable, yet they do not spontaneously decay, much less spontaneously combust in Earth's reactive oxygen atmosphere. We owe our very existence to the universe's many barriers to decay and disorder.

We have proposed that all evolving systems are conceptually equivalent in part because all such systems are subject to selection. But what, exactly, is being selected for? As Google software engineer and artificial life researcher Blaise Agüera y Arcas succinctly puts it, "What persists, exists." Our universe not only generates a remarkable variety of fascinating complex objects, from molecules to crystals to planets to pandas, but those objects also persist long enough for us to enjoy and study them. The nature of persistence, then, must be central to any natural law of evolution. However, persistence is not just one thing. We suggest that persistence comes in three distinct flavors—static, dynamic, and novelty-generating—each of which is subject to selection.

Static Persistence: The most basic selective force is for arrangements of matter—like atoms, molecules, or cells—that don't spontaneously fall apart: what we call static persistence. We have seen that only a tiny fraction of all possible configurations in evolving systems are able to hang around long enough for us to notice them. Such persistence need not be indefinite. Radioactive uranium-238 persists, even though it slowly decays away to the element lead, with half of the uranium atoms transitioning about every 4.5 billion years. Many minerals gradually erode or convert to other kinds of crystals over tens to millions of years—think iron rusting. Biological species also change, some gradually evolving to new species, others going extinct. Nevertheless, all these kinds of objects tend to persist unless some new set of conditions comes along that leads to an even more stable configuration. We call this most basic phenomenon first-order selection. Nature chooses configurations of atoms, molecules, and cells that don't fall apart.

Dynamic Persistence: Persistent systems need not be static and unchanging. Many dynamic systems also persist, but they are not defined by a fixed set of building blocks, such as a mineral's atoms or an atom's protons and neutrons. A star's elemental abundances change continuously over its lifetime. A hurricane incorporates many different parcels of air as it churns across water and land. Living organisms constantly eat and excrete, exchanging matter with their environment. Yet despite the transience of their building blocks, these systems can last for many days to many billions of years.

What does persist is a dynamic system's activities, not its constituents. In accordance with the second law, all dynamic systems require a constant flow of low-entropy "fuel," which is converted into high-entropy "exhaust," to sustain the energetic behavior and coherence of the whole. As long as energy keeps the hurricane winds blowing or the Sun's nuclear fuel burning, the dynamic system persists. When the energy flow ceases—when a star runs out of nuclear fuel or you stop eating—that dynamic system ceases to function.

Other core functions can arise that perpetuate the dynamic system. The ability to replicate is one: The way a wildfire spreads may prolong burning long after the original burned area has cooled off. Likewise, the reproduction of life sustains a species long after many individuals have died.

The ability to self-regulate is another core function. Over

million-year timescales, chemical weathering can act as a planetary thermostat, modifying the atmosphere and damping climate excursions.

And in the most intricate dynamic living systems, forming memories, sensing the environment, and learning to predict changes is the ultimate survival tool. Knowing when the next flood or drought will come can ensure a successful harvest.

We view persistence of such dynamic conditions in stars, hurricanes, life-forms, and many other systems as second-order selection. In these instances, selection operates at the scale of ongoing processes. Among the vast numbers of possible configurations, nature preferentially selects the tiny fraction of arrangements of atoms or molecules or cells that promote the long-term stability of the dynamic system.

Selection for Novelty: The most complicated evolving systems rely on nested networks of smaller complex systems, each level persisting on its own while helping to maintain the persistence of the whole. Life provides the most familiar examples. Cells incorporate many different enzymes—each a specialized molecule that is selected for its ability to perform a vital chemical reaction, such as digesting food or processing genetic information. In other words, an enzyme's function is not to perform any of the cell's core functions in isolation, but rather to play one specific role in the context of a higher level of organization.

From the perspective of an individual enzyme, the selection pressure originates at a much higher level—a level that promotes survival of that lineage of cells. What's more, such selection pressures often exist across many scales. A cell may be an essential part of an organism, which in turn may be an essential part of a community, which in turn is part of a dynamic ecosystem. Selection pressures typically operate at several of these levels simultaneously.

A remarkable feature of such nested complex systems is that completely new functions often emerge. The evolutionary history of multicellular organisms provides many striking examples—eyes to see, fins to swim, legs to walk, and wings to fly all bestowed new abilities and associated advantages for survival. In a universe that hosts a vast range of possible configurations, there is always a benefit to exploring new, highly effective arrangements. The capacity to generate novelty is preferentially selected because it confers a tremendous advantage in a complex and changeable world. We call this process of exploring and exploiting novelty third-order selection.

The discovery and exploitation of completely new functions adds richness and novelty to an evolving system, though such new abilities may become so distant from core functions that they are sometimes difficult to understand in the context of the survival of the larger system. Consider the elaborate dance culture that has been sexually selected for by generations of birds of paradise in Papua New Guinea. Males sport impressive fanlike plumage and engage in dance-like mating behavior with a combination of prolonged vocalizations and intricate feather movements, in some cases with several competitors vying on neighboring perches for the attentions of a female. Such complex displays suggest that selection for novelty is in play, though far removed from the survival of any individual bird.

In a similar vein, in the context of evolving human societies, the creation of art and music might seem to have very little to do with the survival of the species. Nevertheless, their origins likely stem from the critical need to transmit information and create bonds among communities, and such endeavors continue to enrich life in innumerable ways. Like eddies swirling off the main flow of a river, selection pressures for ancillary functions such as language, music, and dance can become so distant from the core functions of their host systems that they can effectively be treated as independently evolving systems, perhaps eventually generating their own core functions.

The ability to continually explore new functions is a hallmark of life. The generation of novelty has the potential to further intertwine the core functions within a nest of feedback loops that promote competitiveness and survival. The invention of eyes to see and wings to fly provided animals new ways to continue performing their core functions, while making multiple lineages of organisms more successful at surviving and reproducing.

Naturalists often see evolutionary arms races among species or groups. Bigger teeth necessitate the evolution of thicker shells, leading to even bigger teeth and stronger jaws. Humans coevolve with a host of pathogenic viruses and cellular organisms. In these cases, and countless others, when two evolving entities compete, they can drive each other to higher and higher levels of sophistication, at times with striking novelty. Multicellularity led to new levels of organization. New modes of locomotion, including flying and walking, opened previously inaccessible environmental niches. Higher levels of information processing ultimately led to consciousness and our ability to search for lawful patterns in the universe. In these ways and many more, the selection pressures of a constantly changing environment breed an incessant drive toward novelty.

Life also has proven adept at finding surprising new uses for old functions—a process called exaptation. The wings of insects may have originally helped to regulate body temperature. Similarly, feathers on dinosaurs may have originally performed thermoregulatory, display, and biomechanical support functions before aiding in flight. And, in a twist on the exaptation theme, penguin wings once used for flight are now repurposed as fins for efficient swimming in cold polar climes. This concept of exaptation highlights the importance of *context* in the evolution of ancillary functions. Heat regulation tends to favor high-surface-area structures that can be flapped to generate cooling air currents. However, the exaptation for flight is not preordained—think elephant ears; they play a key role in moderating an elephant's temperature but are in no way suited to modification for flight (though Dumbo the flying elephant is a delightful concept!).

Art, literature, music, games, and technology in human culture are all examples of third-order selection. Each pursuit reflects our inherent desire to experiment with our world to discover new ways of thinking, being, and communing with one another. Although it may be argued that humanlike innovation has negative adaptive value, as evinced by flirtations with self-inflicted collapse, so far, our evolutionary success as a species may be attributed, in large part, to our curiosity. Perhaps it will be humanity's ability to learn, invent, and adopt new collective modes of being that will lead to its long-term persistence as a planetary phenomenon.

In light of these considerations, we suspect that general principles of selection and function, which are clearly in play in numerous natural chemical systems, may also apply to the evolution of symbolic, technological, and social systems—for example, the exaptation of an oxcart's wheel by a clever potter, the conversion of a hunter's bow into a musical instrument, or the development of computer code from written language.

Selection and Function Are Contextual

We've reached a point in our argument where confusion and doubts may arise. Isn't this discussion of selection all a kind of tautology? Things persist because they are able to persist. Things that learn better ways to persist will persist better.

In one limited sense, that's exactly what we are saying. Evolution occurs in all kinds of systems because some configurations are better than others at not falling apart. But it's not quite that simple because selection for function carries a meta-question. What aspect of a system is persisting, and what exactly is the function being selected? In the case of atoms and minerals, static persistence seems to be the most obvious function. But how do we identify relevant functions in the context of complex evolving systems, especially life?

In the case of an enzyme—a biomolecule that promotes a specific chemical reaction in a cell—the problem is often straightforward. Consider nitrogenase, an enzyme that helps to convert atmospheric nitrogen gas into biologically essential ammonia. There exist numerous variants of this molecule, some more efficient than others. We can measure the efficiency of a given nitrogenase molecule in the lab: The faster the rate of ammonia production, the greater the function of the enzyme.

The concept of function becomes much more nuanced when considering evolving systems with nested operations. Every cell displays layers of function—some at the level of small molecules, some at the scale of internal cellular structures, and others related to the individual cell or even communities of many cells. By converting nitrogen into ammonia, the nitrogenase enzyme serves the persistence of its host cell, but it also plays an important role in maintaining the global nitrogen cycle. Hence, one can explain the proliferation of nitrogenase on Earth by how it promotes the survival of individual cells, or by how it helps maintain the functionality of entire ecosystems. Both are right, and the answer you choose to emphasize depends on the scale you're interested in.

We have suggested that nature selects first and foremost for persistence—for survival. At the scale of the individual atom or the isolated mineral species that may be the most logical conclusion. But in a more complex, intertwined, evolving system, such as a diverse ecosystem or the coevolving geosphere and biosphere, the answers become much more nuanced. The death of an individual cancer cell may promote the persistence of an organism. The death of an individual human has minimal effects on the persistence of the species.

In some fascinating instances, critical configurations are not, themselves, stable. Function in those cases depends on being stable *just long enough*. The nucleosynthesis of carbon-12 from three helium-4 nuclei is a good example. Carbon-12 doesn't form by the highly improbable simultaneous collision of three helium-4 isotopes. Rather, it's a two-step process with a pair of helium-4 nuclei fusing first to make beryllium-8. The problem is that beryllium-8 is extremely unstable, breaking apart on average in about a quadrillionth of a second. Fortunately, a significant fraction of those ephemeral beryllium-8 isotopes in stars fuse with a third helium-4 before that disintegration happens. The result is stable carbon-12, and all the novelty that follows (including the origins and evolution of life). In that scenario, when persistence and stability are lacking, one might justifiably ask: What is the function of the transient configuration called beryllium-8?

Before life emerged, an individual phosphate mineral's most obvious "function" was to persist as a crystal, but in the larger context of our evolving, living planet the function of that same mineral might be to hang around just long enough to dissolve and release essential phosphorus into the biosphere. Similarly, basalt is the durable first rock to cover the surfaces on almost all our solar system's rocky planets and moons. In a sense, its primordial function was to solidify and persist. But today on Earth basalt is the essential precursor to rich volcanic soils, vital to agriculture in many parts of the world, while it plays crucial roles in soaking up excess carbon dioxide from the atmosphere. Thus, in the context of the coevolving geosphere and biosphere, basalt plays many roles beyond mere persistence. Evolution of the biosphere as we know it would not have been possible without a *lack of persistence*—that is, the destruction by weathering—of vast swaths of basaltic crust.

The conclusion must be that selection for function is often contextual. If you are a farmer or a stonemason or an ecologist or an artist, your concept of basalt's function will differ. This situation is quite different from many other less subjective scientific measurements or calculations. The mass and chemical composition of an enzyme, for example, are independent of what function we are considering.

The three sources of selection—static persistence, dynamic persistence, and novelty generation—are not always equally prominent, nor do their relative weights stay steady through time. Which selection mechanism dominates in any given scenario will shape the final result. When selection mechanisms change, the system can take on a totally different flavor.

Consider a time before life emerged on Earth. Complex organic molecules flit in and out of existence. A nucleic acid might form on a clay mineral at one moment, then get shredded by an ultraviolet photon the next. In this chemical soup, static persistence dominates the question of what survives and what does not. Molecules that are the most resistant to change stick around the longest.

But one day, some *collection* of organic molecules finds itself in a fortuitous arrangement. These molecules begin to interact in a way that promotes their mutual persistence. Perhaps A transforms into B, which reacts to form C, whose destruction by ultraviolet light forms A. Draw a picture in your mind: A forms B, B forms C, and C closes the loop by forming A—a cycle! Although C is fragile, falling apart in the presence of UV radiation, it will always be present because it serves the function of helping to restock A: The more A is used up to create B, the more B will make C, and the more C will regenerate A. Molecule C has been selected not because it is a master of static persistence but because of its function in maintaining a dynamically persistent system composed of A, B, and C. In this system, dynamic persistence has usurped static persistence as the dominant sculptor of what molecules persist. Something akin to this simplistic illustration likely occurred along the way to life.

■

Such musings raise a deeper question. A view of the natural world that highlights selection for function invites us to speak of meaning, value, and perhaps even purpose. A certain molecule may be valuable to a chemical cycle's persistence. A certain gene might serve a purpose in an organism. A certain geologic cycle might mean something to the stability of a planet's climate. Scientists typically remain neutral on questions of value and purpose. Yet consider the obvious fact that some complex arrangements of organic molecules form highly functional systems (e.g., you), while other equally intricate arrangements of molecules do not. In this context, are not certain evolving systems in some deep and fundamental way more . . . What is the right word? Fascinating? Dynamic? Versatile? Important?

A functional perspective of nature has the potential to change the way we understand the story of the universe—and our place in it. We'll explore these implications in the final chapter. But before we get there, we must address the thermodynamic elephant in the room: Why isn't the law of increasing entropy and its vaunted arrow of time enough?

4

Entropy Versus Information

Some deeply thoughtful and informed scholars disagree with the central premise of this book. They argue that diverse evolving systems do not require a new explanatory and predictive law of nature—especially a law coupled with a second arrow of time. The inexorable increase in entropy is all you need to explain evolution, they say. Are they right?

Isn't Entropy Enough?

The second law of thermodynamics applies everywhere, all the time. Any shuffling of atoms, no matter how subtle, any flow of energy, no matter how trivial, indeed any system that changes in any way at all will experience a dissipation of some energy and an increase in entropy. The second law of thermodynamics is integral to any change in energy from one form to another—in a nuclear fusion reaction, in a transfer of heat energy, or when you eat your vegetables. Anytime you take a walk, or hug a child, or simply think about walking or hugging, the entropy of the universe increases ever so slightly. That is the law.

What we think is missing in the laws of energy is that tricky concept of *selection for function*. The law of increasing entropy applies to all possible configurations with equal weight, but nature embraces some configurations while rejecting others. Yes, of course, the second law of thermodynamics is in play in any evolving system whenever selection occurs. In the case of static persistence—such as the formation of any given stable atom from protons and neutrons or the formation of any given stable mineral species from chemical elements—the reactions can be, at least in principle, completely described, explained, and predicted by the laws of thermodynamics. The protons and neutrons (in the case of atoms) or chemical elements (in minerals) seek states of lower energy. In the process of forming atoms or crystals, heat energy is released to the surroundings, and the total entropy of the system increases. Even when a seemingly disordered state like salt water evaporates to produce beautifully ordered salt crystals, the laws of thermodynamics are in control. The increase in local order that results from the formation of each individual salt crystal is more than compensated by the inevitable waste heat energy that radiates out into the vastness of space.

An additional wrinkle applies to complex evolving systems. Our descriptions of atom evolution in stars and mineral evolution on Earth add the idea of a sequence of processes that leads to increasing diversity—a kind of bootstrapping from simpler beginnings to ever more complex states. Those evolutionary details don't always seem to flow inevitably from the second law. Nevertheless, every individual step of those pathways is guided by the laws of thermodynamics. Consequently, we will concede that a new law of evolving systems is not essential to describe and explain many examples of static persistence.

Unlike static persistence, which results in energy dissipation only during the formation of a stable atom or mineral, dynamic persistence requires active, ongoing dissipation of energy and an associated continuous increase in entropy. Such continuous transfers of energy might be challenging to model. Nevertheless, once again it seems possible in principle to describe, explain, and predict the behaviors of a star or a hurricane based on the existing laws of nature, including the laws of thermodynamics. Each tiny part of these systems, down to the submicroscopic interactions among protons or atoms, is rigorously described by those powerful laws.

But here's the key point. While dissipation of energy must accompany any change in the evolving system, that dispersion of energy into space and the accompanying increase in entropy cannot completely describe, much less explain, how many complex evolving systems with layer upon successive layer of novelty come into being in the first place. Nor can those laws predict how they will change going forward. We suggest that something else, some other natural process, must participate in evolution.

■

Life illustrates this point. Every chemical reaction in a cell obeys the laws of thermodynamics. They must. Nevertheless, the laws of thermodynamics do not describe and explain, much less predict, why life emerged from the primordial soup. Many possible nonliving configurations of atoms and molecules could equally well satisfy the laws of thermodynamics. In fact, many mixtures of small molecules are much more thermodynamically stable than the complex molecular components of a cell, which is why dead bodies quickly decay to smelly gases.

The transition from nonliving molecules to a living cell is where selection for novelty must come into play. Some configurations of molecules are selected because they are better at making copies of themselves than others. The rejection of some configurations is not necessarily because that arrangement is unstable, or even because it doesn't "work." One viable configuration is often rejected simply because another competing configuration performs the same critical function better.

We suggest that the choice of one configuration over another is not a matter of thermodynamics. Entropy, alone, is not enough to explain evolution. Rather, the behavior of complex evolving systems is rooted in another fundamental aspect of every complex system—*information.*

What Is Information?

We live in an age when new information-based technologies often dominate the news. Bitcoin mining, genetic algorithms, artificial intelligence, and much more are rooted in the science and engineering of information. We contend that information also provides the best way to describe and explain evolving systems. But what, exactly, is information?

In the most basic sense, information is any kind of description that can be reduced to a simple string of zeros and ones—what are known as binary digits, or bits. That means information can be quantified and manipulated in a digital computer. Music and videos can be converted to digital recordings. Photographs can be recast into the digital form of pixels, each a tiny colored square. Any string of numbers or letters, similarly, can be reduced to zeros and ones. But what is it that is being quantified?

No one answer is possible, because "information" means different things in different contexts to different people. Colloquially, information just means what we know about a topic—your address, cell phone number, and date of birth are basic pieces of information about you. Information of this sort often comes in tables or spreadsheets—think of an old-fashioned telephone book or a library catalog.

More technical definitions of information may involve a sequence or arrangement of components to provide a complete description of a system. The precise arrangement of atoms in a crystal, the sequence of DNA letters in a gene, or a sequence of dots and dashes in a Morse code message represent critical information required to describe those systems. As systems become more complex, the nature of information may take many forms: for example, music videos, maps of Earth in three dimensions, and more abstract systems in much higher dimensions—think about a network that represents all the interactions among molecules in Earth's atmosphere, or all the cell phone calls made yesterday around the globe.

Given this diversity of applications and approaches, there is no one universal definition of "information." Even so, a few useful kinds of information point us in the right direction. Perhaps the most basic measure of the information contained in any object or message is Kolmogorov complexity, which is defined as the minimum amount of information (in bits) required to exactly reproduce that object or message. For example, a sequence of 100 DNA letters, in which each of the 100 letters has four different possible choices, offers 4 to the 100th power (that is, 4 multiplied times itself 100 times: 4 × 4 × 4 . . .) different possible configurations. That's equal to

2 to the 200th power, or 200 bits, of information. As long as the string of letters is random (that is, with no repeated sets of the same letter sequence or other internal symmetries), then all such DNA strands have a Kolmogorov complexity of 200 bits, regardless of whether the strand does anything useful or not. The same is true of a computer program with 1,000 similar lines of code. Whether or not the program works, the Kolmogorov complexity is the same.

We now understand that messages in any format, including letters, numbers, Morse code, photographs, a microbial genome, and more can be reduced to a sequence of zeros and ones—the binary digits of information theory. The application of this idea is relatively straightforward *as long as you don't think about context and function*.

Kolmogorov complexity is a quantitative, scientific measurement that can be applied to any system with large numbers of different possible configurations, without any thought to value or context. Function doesn't matter. Yet, when considering the role of information in complex evolving systems, the function and the context are critical.

Information and Function

How might we understand the possible connection between function and information? Consider a simple example from the world of computer coding. Let's say you have an elegant 1,000-line program that is supposed to calculate which mineral species is most stable under certain conditions of temperature, pressure, and composition—a computer program based on those elegant laws of thermodynamics. But your code doesn't work. There's one little bug on one line of code—a missing

")". Admit it. Whether you've ever written code or not, it's the kind of mistake that has happened to all of us.

So, here's the point. You can mistakenly insert a "]" for the missing ")" and the program still won't work. In spite of this error, by typing in that one character "]" you have used a tiny bit of energy, which causes a tinier bit of energy to radiate out into space and the entropy of the entire Earth–Sun system to increase by a minuscule amount. The second law has spoken, as it must.

On the other hand, you can correctly type the missing ")" and the program suddenly functions just the way you hoped. By typing in that one character ")" you have used exactly the same tiny bit of energy, which causes the same tinier bit of energy to radiate out into space and the entropy of the entire Earth–Sun system to increase by the exact same minuscule amount. In each case—whether you type correctly or not—the laws of thermodynamics have been obeyed, as they must.

But there is a critical difference between these two examples. In one case the resulting configuration is useless, while in the other case you have a functioning computer program. In one case the information embedded in your computer program is useful, in the other case useless. Naturally, you will select the code that works properly—code that might be further modified as it evolves into better and better versions.

The same logic applies to information systems in biology, which are based on the genetic code of the double-helix DNA molecule, with its four letters A, T, G, and C instead of zeros and ones. Imagine a gene—a long strand of DNA that is supposed to code for a critical enzyme that digests proteins in a cell. However, a mutation has led to a mistake in one letter—perhaps an errant T replacing the correct G—and so the enzyme doesn't work. A cellular mechanism tries to correct the defect, replacing the incorrect T with an equally incorrect C. The resulting modified enzyme is useless, but in that process of chemical substitution a minute bit of energy is used to break and reform chemical bonds, some even smaller amount of energy ultimately is lost to space, while entropy increases ever so slightly. The second law of thermodynamics has been satisfied, as it must.

Now, retry the chemical replacement with the correct letter G. The same minute bit of energy is used, the same smaller amount of energy is ultimately lost to space, and entropy increases ever so slightly. But in this case the defective enzyme works. The cell will survive because the information embedded in DNA yields a functional molecule. Evolution will select the DNA sequence that works, so the cell can survive and continue to evolve.

Inevitably, in these processes some energy disperses and entropy increases; the second law of thermodynamics always plays its role. But energy and entropy are not what's most important in the survival of a cell. There's nothing in the roles of energy and entropy that distinguishes the correct DNA code that works from the faulty DNA code that doesn't. The selection of one DNA strand in preference to another must arise by a different process—a process that is not dependent on increasing entropy, but rather increasing *information*. Information, not entropy, must guide any selection process that favors one arrangement of atoms, or computer code, or DNA letters over others.

But here's the problem. Kolmogorov complexity cannot help us. Whether the system works or not, the Kolmogorov complexity of a random DNA sequence is the same. In reality, because the language of DNA is written in three-letter-long "words," the Kolmogorov complexity of a functional DNA strand is generally a little *lower* than that of gibberish DNA. Because of internal repeats and other symmetries, the patterns of triplets in functional DNA can be described with less Kolmogorov information than you would need for a completely random DNA string of equal length. But shouldn't a measure of complexity comport with our intuition that living things are a lot more complex than junk? Consequently, we need a different measure of information—one that includes the idea of a system's function.

Functional Information

Context is critical when thinking about the information content of a complex system. Consider three states of a chicken—a live chicken, a dead chicken, and a chicken egg. All three states have the same genetic information; the complete sequence of DNA letters that stores the requisite information to "make" a chicken is present as strands of DNA in each state. However, only the live chicken and dead chicken possess structural complexity—the kind of information related to appearance and internal anatomy that distinguishes a chicken from other animals. And only the live chicken displays the behavioral complexity of a clucking, pecking, strutting chicken—a different context to think about when it comes to information. Which point of view is correct when thinking about the information of a chicken—genomic, anatomical, or behavioral? We suggest that there is no one correct answer. It depends on the context of your inquiry. What function is important to you?

In 2003, Harvard biology professor Jack Szostak (b. 1952), subsequently cowinner of the 2009 Nobel Prize for Physiology or Medicine, introduced a brilliantly simple solution to the problem of combining concepts of information and function. He called his alternative flavor of information "functional information." His one-page paper in the prestigious journal *Nature* specifically addressed an application of information to biopolymers such as DNA (the famed double helix) and RNA (a single-stranded cousin of DNA). As we have seen, these long strings of molecules contain information by incorporating sequences of four different molecular letters. Any given position in one of these molecules can be defined with two bits of information—perhaps 00 for A, 01 for C, 10 for G, and 11 for T. Any random strand with 100 RNA or DNA letters has a Kolmogorov complexity of 200 bits of information, whether or not that strand does anything useful.

Szostak's insight was that only a tiny fraction of all possible sequences does something useful—has a function. In his case, he had been studying strands of RNA called aptamers that are able to bind strongly to a second specific molecule by wrapping themselves like a glove around that other molecule. A 100-letter RNA strand has 4 to the 100th power different possible configurations—an immense number of possible arrangements (in fact, a 16 followed by 59 zeros!). Szostak knew that the vast majority of RNA strands have no ability whatsoever to bind in this way. In the context of binding, those useless molecules are functionless even though their Kolmogorov complexities are a whopping 200 bits. On the other hand, a tiny fraction of that astronomical number of possibilities can bind just a little bit to the target molecule. And if you insist on stronger and stronger binding (remember, in this case, binding to a target molecule is the desired function in Szostak's experiments), then a tinier and tinier fraction meets the demands.

Technically, functional information is defined as the negative logarithm to the base 2 ($-\log_2$) of the fraction of configurations that can achieve a desired function. The $-\log_2$ function turns that fraction into bits, a standard unit of information. Understanding the mathematical details isn't critical to our argument. All you need to know is that the *fewer* configurations that can perform a given function—that is, the more exceptional the arrangement—the *higher* the functional information of that system.

Functional information is a measure of the probability that a sequence of RNA will achieve a desired degree of function. In the case of RNA strands that are exactly 100 letters long, if only one sequence yields the desired result, then the functional information of that unique, special sequence is the maximum possible 200 bits—the same as the Kolmogorov complexity of a random sequence. However, if many different configurations achieve a functional objective, for example, if millions of strands can bind to Szostak's target molecule, then the functional information will be less than 200 bits. The easier it is to achieve a function, the smaller the functional information.

Consider three examples. First, most strands don't bind at all, so their functional information is zero—no function equals no functional information, even if the Kolmogorov complexity is 200 bits.

Alternatively, if 1 in every 100 configurations binds just a tiny bit, then the functional information of those most weakly bound RNA strands is a modest 7 bits (i.e., the negative log to the base 2 of 1/100). That's because the functional information of an individual strand is calculated based on what fraction of other strands out of the total range of possibilities "worked" as well.

A third example is represented by one set of experiments in which Szostak found that one configuration in every 100,000,000,000 achieved a moderately high binding strength.

Any one of those many functional RNA strands has a greater functional information of 37 bits (again, the negative log to the base 2 of 1/100,000,000,000).

A few years later, one of us (Bob) teamed up with Szostak to explore wider applications and implications of the functional information concept. For example, we considered language—specifically the case when combinations of letters attempt to convey a critical message about the location of a house fire. Most letter combinations are garbled nonsense, and others have unfortunate typos or are in a foreign language, while only a tiny fraction of letter sequences convey the urgent message. Relatively few combinations of letters—compared to the oodles of combinations that could possibly exist—convey actual meaning. So, the functional information of a cursory emergency message is large, while the functional information of your favorite book is enormous.

We also examined sequences of computer code that simulated lifelike replication of its own code. Again, most random sequences of code are useless with no functional information, while only the tiniest fraction of possible code sequences displayed the desired function. Consequently, the functional information of self-replicating code is quite large as well. We left the door open for applications to many other systems, including artificial evolving systems, such as the suite of computational tools called genetic algorithms, and possibly life, as well.

The Conundrum of Function

Functional information offers a powerful way to compare quantitatively evolving systems in which only a tiny fraction of all possible configurations displays a desired function, such as static persistence, dynamic persistence, or novelty generation. In any such system, most configurations will have no function whatsoever, while only a tiny fraction will be functional to one extent or another.

But there's a conundrum. Most physical attributes—the mass and chemical composition of an RNA molecule, for example—have a fixed value that is not dependent on context. It doesn't matter if the RNA molecule is in a cell, in a test tube, or in the vacuum of space—its mass and composition are constant. Calculating functional information is different because you first must choose what function is of interest. An RNA aptamer that can bind well to one specific molecule might not interact at all with another equally significant molecule, much less perform an essential cellular function like digesting food or making proteins.

Consider an everyday example. In the United States, an urgent message in English will likely get a quick response, while the same message in Chinese will usually be met with blank stares. By contrast, in China the opposite is true. The Kolmogorov complexity of the two messages in English and Chinese might be exactly the same, independent of the location in which they are spoken, but the functional information is dependent on geographic context. This aspect of information confronts us all the time. An eloquent speech is effective only if you speak the language. A compact disc of your favorite music is useless unless you have a CD player. An RNA strand that binds perfectly to a target molecule does nothing if there are no target molecules with which to bind. The eloquent speech, favorite CD, and RNA molecule all possess lots of information, but that information is useless in the wrong context.

Functional information (as opposed to Kolmogorov complexity) is context dependent. We, as observers, must decide for each context which functions are important and which ones aren't. We must, in essence, assign relative values to different states of an evolving system, and that task seems subjective—perhaps even unscientific.

But what does nature tell us? As we have seen, numerous natural evolving systems display three diagnostic traits: (1) the potential to adopt numerous different configurations, (2) processes to generate many of those configurations, and (3) preferential selection for configurations that work better. Nature shows us through the mechanism of evolution that some configurations are favored over others—they are selected among the vast number of possibilities because of their functionality. In other words, some configurations have greater functional information, and those configurations will be preferentially selected. This selection process is most assuredly context dependent. That is the way nature works, and the inexorable drive to increase functional information is the key to describing, explaining, and quantifying that selection process.

Increasing Disorder Versus Increasing Order

We began this chapter by asking, "Isn't entropy enough?" Isn't the drive to increase entropy sufficient to describe and explain how evolving systems work? We suggest that the answer is no, and here's why. In a selective environment, both entropy *and* functional information must increase, but these two physical quantities are different and, under many circumstances, decoupled from each other.

The increase in entropy and disorder is one arrow of time.

Any change in an evolving system will be accompanied by some movement of energy, some loss of energy to space, and some increase in the total entropy of the universe.

Nothing in that statement implies a change in the system's ability to persist. In an evolving system subject to selection for function, it is functional information, not entropy, that describes and explains how the system will change. And here we glimpse a possible path forward in our quest for a missing law of evolution: What if evolution obeys a law of increasing functional information? That increase would represent a second arrow of time—an arrow of increasing localized order and complexity. If our conjecture is correct, then time's second arrow is completely consistent with entropy's increasing universal disorder, but it is a different and independent phenomenon.

Quantifying Evolution

In our quest for a natural law of evolving systems—systems that evolve because of their propensity to select for advantageous functions—functional information is a powerful metric. Each time a selection event occurs in an evolving system, whether producing a new stable configuration of protons and neutrons to make a novel element or a new arrangement of cells to create a novel life-form, the function of the system must improve. That means, by its very definition, functional information *with respect to that selected function* must increase.

All macroscopic laws of nature share an important characteristic. They all describe quantifiable natural processes in terms of real physical variables. Newton's laws of motion employ mass in grams and acceleration in meters-per-second-per-second to quantify force in a unit called newtons. The gravitational and electromagnetic forces are examples of real phenomena measured in newtons. Similarly, the laws of thermodynamics deal with energy measured in joules, though in the case of energy more than a dozen different units confuse the specialized technical literature—ergs, therms, BTUs, kilowatt-hours, reciprocal centimeters, foot-pounds, calories, and more—each common to energy used in its specific scientific or industrial context. But what unit of measurement could possibly quantify an evolving system, much less unify all the varied examples nature has to offer?

Scientists have thought about this question for many decades, and a few proposals have been offered, most in the context of energy. One recurrent idea is that evolution is tied to the movement of energy. According to this view, evolving systems emerge from nature's imperative to dissipate energy—to increase entropy. It's a subtle idea, and one to which we don't subscribe, but highly evolved, patterned systems do often seem to throw off a lot of waste heat. Your brain is a prime example—it uses a remarkable amount of energy, perhaps 20 percent of every calorie you consume, and in the process radiates a lot of heat into your surroundings. That means the brain accomplishes something ordinary matter cannot. It increases the transformation of energy from calorie-rich food on your dinner table to waste heat radiating out into space much faster than would otherwise take place. In that context, maybe evolving systems can be measured by the units of energy, or perhaps the rate of energy flow.

By contrast, we suggest the answer to evolution's quantifiable unit is information. (Ask yourself, *why* do brains consume all those calories? To process information!) Evolution can be measured with the standard unit of the information age—binary digits, or bits. Most scientific units of measurement quantify something tangible, something you can touch or feel. Mass has weight. We experience acceleration. Forces are in play every moment of our lives. Even as abstract a quantity as energy has real everyday manifestations: A well-aimed snowball hurts. We pay for a tankful of gas. And that donut you just ate has more calories than you'd have wished (though, yum!).

At first blush, information seems very different. How can one look at a natural evolving system and measure it in terms of bits? You can't hold bits in your hand or feel their weight. Nevertheless, our everyday experiences, especially in our robustly digital age, suggest that bits are real. Like energy, you have to pay for information in varied forms—internet access, streaming services, and your favorite music app, all priced by the bit (or in some cases the byte, which equals eight bits). We have to pay to store bits of information on flash drives or in the cloud; storing more bits costs more money. And we all have the intuitive sense that some objects hold more information than others, even if they have the same mass. One pound of pure water would seem to hold less information (in bits) than a one-pound loaf of bread, which in turn holds less information than your one-pound laptop computer or pet guinea pig.

The Law of Increasing Functional Information

The compelling concept of functional information leads us to an important characteristic of every evolving system—what we suggest is a universal law of evolving systems:

The functional information of a system will increase (i.e., the system will evolve) if many different configurations of the system are subjected to selection for one or more functions.

This simple, intuitive idea applies to the widest range of physical, biological, and symbolic systems. Each evolving system has the potential to exist in astronomical numbers of potential configurations, yet each successful system displays only a small fraction of those possibilities. Any process that selects from a population of configurations based on improved degrees of function will inevitably result in increased functional information with respect to that selected function. In other words, the system will evolve.

Like any law of nature, this statement carries with it many implications, not to mention unanswered questions and avenues for further research. And so, as with any new idea in science, we have to ask what predictions might we make, and how would we know if we're wrong?

5

Testing the Law

A universal natural law is much more than a formal statement that organizes and describes observations recorded and experiences remembered. If properly formulated and applied, a robust natural law will lead to new insights about what we observe, as well as explicit, testable predictions about the natural world.

As far as we can tell, the ten natural laws we list in Chapter 1 accurately represent the behavior of macroscopic phenomena, from rainbows to planetary orbits to steam engines. These ten statements are regarded as laws because they have all outlasted repeated testing and scrutiny. In other words, they are survivors. Over the millennia, many other scientific propositions have fallen by the wayside. We no longer speak of light propagating through a "luminiferous aether" or heat being carried by a self-repellent "caloric" fluid. We do not believe, as Ptolemy did, that everything in the solar system orbits Earth. Our periodic table is not home to only 4 elements—earth, water, fire, and air—but 118 (and counting).

It would seem that science itself progresses through the universal principles of evolution we have outlined: Many hypotheses are put forward, but few persist. How can we convince others and ourselves that the law we are currently proposing—the law of increasing functional information—is a keeper? The answer is implied by the law itself: To be selected, it must perform the functions of a natural law. That is, it must be challenged in the most rigorous ways possible, with quantifiable predictions that have been tested over and over again. Even if every one of these tests is passed, the law can never be definitively proven. There's always another observation or experiment that might prove us wrong. Therefore, with each new successful test, scientists can only assert that it is less likely to be wrong. In the meantime, each verified prediction brings us a little closer to understanding pervasive patterns in an evolving universe.

In a very real sense, the law of increasing functional information has been tested and demonstrated for billions of years. Atoms and isotopes have evolved. Molecules have evolved. Minerals and life have evolved, eventually to be followed by evolving language, mathematics, music, and scientific understanding itself. Equally important, the law of increasing functional information may help to explain some recent discoveries, while making solid, testable predictions of others. Here are a few of them.

Aptamer Evolution

The evolution of RNA aptamers is perhaps the most compelling laboratory demonstration of the law of increasing functional information. Indeed, it was RNA aptamer experiments that led Jack Szostak to propose the concept of functional information in the first place. The experimental protocols, though technically difficult, were conceptually simple and elegant.

The initial experimental setup involved two chemical ingredients. Researchers prepared a solution with legions of different RNA strands—in some trials, 100 trillion different arrangements, each with 100 molecular letters. That's a lot of configurations to test (an essential trait of evolving systems), but only a minuscule fraction of the 10 to the 60th power (10 multiplied times itself 60 times) possible combinations of a 100-letter RNA strand (another essential trait of evolving systems).

The other part of the chemical setup was a beaker that held carefully prepared tiny polystyrene beads, each one decorated with numerous molecules called GTP that dangled outward like molecular fuzz. GTP molecules have a distinctive shape that invites another molecule to bind tightly, not unlike a glove enveloping a hand. Only a precisely shaped molecular glove will do the trick—an ability measured by a quantity called binding energy. The challenge is to design an RNA strand of 100 letters so that it is the perfect "glove" to GTP—a strand that has the maximum possible binding energy.

Jack Szostak knew that most RNA strands have no binding affinity to GTP whatsoever. In the language of our model, they have zero bits of functional information (even though the Kolmogorov complexity of a random 100-letter strand is 200 bits). Pour those useless strands into the beaker with GTP-coated beads and nothing happens. However, with a huge population of 100 trillion different RNA strands, a small fraction will stick, albeit poorly, to the GTP.

With that context, the first step of the experiment involves nothing more than pouring the first solution with 100 trillion random RNA strands into the beaker with GTP-coated beads and waiting a little while. Pour out the solution, which still holds the vast majority of useless RNA strands. Then treat the beads with a bit of chemistry that releases the tiny fraction of RNA strands that were still clinging to GTP, however weakly. After that first experimental cycle Szostak and his team had concentrated perhaps a few thousand RNA strands that possessed some small degree of binding energy to GTP.

The next step required a nifty trick invented by biochemists some years earlier. They copied those few surviving RNA strands over and over again, but they didn't copy them very accurately. In this process, they synthesized a new population of 100 trillion strands, each one a mutated variant of a first-generation strand that had stuck just a little to GTP. Through these accidental mutations, the population of RNA strands explored new configurational possibilities.

Then they repeated the whole process—a second round of experiment with new GTP-coated polystyrene beads and their second generation of slightly evolved RNA strands. As you might expect, some of those evolved RNA strands stuck a little bit better, with greater binding energy, than anything from the first round. This evolved fraction had increased functional information.

Rinse and repeat. They synthesized a third generation of 100 trillion evolved RNA strands, then a fourth and a fifth. With remarkable efficiency, each generation achieved significantly greater binding energy by finding better and better solutions to the hand-in-glove challenge. As the floor for being a "good binder" rose, the RNA strands that were successful represented a tinier and tinier fraction of all possible configurations. Recall that functional information increases when the fraction of functional configurations decreases. In just a few generations, the functional information of the best binders soared; each tenfold increase in binding energy represented an additional 10 bits of functional information.

The entire process required no more than seven or eight cycles—just a few days of lab work. And yet, at the end, Szostak's brilliant experiment had produced RNA strands that literally fit like a glove—close to the theoretical maximum binding energy.

The aptamer evolution experiment of Szostak's lab is a stunning example of the law of increasing functional information. Start with an RNA-based system that possesses inherent combinatorial richness. Generate large numbers of different configurations. Subject those configurations to selective pressures. Repeat the selection process and the system evolves. Step-by-step, the functional information increases.

Genetic Algorithms

One of the strongest arguments that evolution is inextricably linked to information, as opposed to energy dissipation, is the observation of evolution in purely symbolic contexts. Languages, mathematics, computer code, and scientific knowledge all evolve because these systems share three critical characteristics: (1) they rely on symbols that have the potential to be arranged in vast numbers of different ways, (2) many different arrangements are generated and tested, whether by design or random shuffling, and (3) selection processes favor some configurations that display useful functions. Among the countless examples of evolution in purely symbolic systems, the class of computer programs called genetic algorithms demonstrate astonishing evolutionary processes using nothing but a computer and some clever code.

The prime targets of such calculations are the designs of physical objects that are virtually impossible to invent on the lab bench. A classic, jaw-dropping example was published in 1997 by computer scientist John Koza of Stanford University and four colleagues. Their task was to use genetic algorithms—populations of candidate solutions to a real-world problem written in computer code, which compete and evolve in silico—to design optimal electrical circuits with varied desired inputs and outputs, such as an amplifier and a noise filter. Their most difficult challenge was a so-called computational circuit in which the output voltage equals the cube root of the variable input voltage (i.e., a numerical value for the output voltage that can be multiplied by itself 3 times to yield the numerical value of the input voltage). Give a hundred electrical engineers a hundred years, and they would have difficulty achieving that objective by hand calculations and benchtop fiddling.

In a nutshell, Koza and colleagues threw a bunch of electronic components—dozens of varied diodes, transistors, resistors, and capacitors—into a box, where the computer program generated numerous randomly interconnected, three-dimensional virtual circuits of bewildering complexity. Thanks to the rigorous laws governing electrical circuits, the computer could calculate the input/output characteristics of each circuit and identify a few circuits that very roughly matched the desired result.

The next step in any genetic algorithm is to take the best results from round one, virtually mutate them by randomly breaking some connections and building others, and then rerun the calculations. As in Jack Szostak's aptamer evolution experiments, each cycle of computer evolution yielded a better solution with increased functional information. After a few cycles, this method converged to an optimal solution—a result taking seconds of computer time.

One of our favorite examples of a genetic algorithm is NASA's ST5 spacecraft antenna, which was programmed in 2006 by a team of computer scientists at the storied Jet Propulsion Laboratory in Pasadena, California—the place where many of NASA's spacecraft are engineered and built. The challenge was to design a lightweight, power-efficient antenna with strict requirements related to bandwidth and "beamwidth"—an antenna that could fly on a spacecraft and send signals back to Earth from deep space. Using an evolutionary algorithm that mimics Darwinian evolution, the program first generated large numbers of relatively simple antenna designs (i.e., large numbers of configurations), each based on a short, bent wire. Each of the computer-generated designs was evaluated for its efficiency (i.e., selection). This selection step was possible because the behavior of antennas can be calculated by sophisticated application of Maxwell's electromagnetic equations.

In a second cycle of evolution, the most successful designs from the first round were modified by the program by randomly changing the lengths and kink patterns. Each new antenna from this second generation of designs was evaluated and the best designs selected for additional cycles of mutations and selection until an optimal design was generated. The resulting antenna—a bent wire about three inches long with a bizarre series of five random-looking kinks—far exceeded the capabilities of any product of manual design. Proof of concept followed when the ST5 antenna was flown side-by-side with the best engineered design. The evolved antenna won the contest by a wide margin.

Genetic algorithms are now used extensively in science and

industry to solve all manner of complex problems from the shapes of airplane wings to the design of new drugs to the solution of sudoku puzzles. In all instances, these applications follow the law of increasing functional information: With each cycle of mutation and selection, a smaller and smaller fraction of all possible configurations achieves higher and higher function. The functional information increases.

The lure of genetic algorithms, in contrast to evolving systems of atoms, minerals, or life, is that the computer can generate and test millions of new configurations every second. As a consequence, computer-driven evolution offers remarkable new opportunities in science and technology for rapid discovery.

At the same time, if an artificial intelligence system can evolve at the rate of a million generations per second, are there limits to its potential? Is computer evolution open-ended and, if so, what unexpected, novel capabilities might such a system display?

As we write this book, humanity stands at the dawn of a new age of generative AI. Publicly available large language models (or LLMs) can create any body of text, from English essays to political satire, in a matter of seconds. Trained on an enormous corpus of data mined from the annals of the internet, LLMs can also solve crossword puzzles, answer math questions, write computer code, pass the bar exam, and hold romantic conversations. Similar tools abound for generating original audio, images, and videos at users' beck and call. Will these astonishing programs be a net benefit or detriment to humanity? Experts argue both sides.

Through the lens of the missing law, we gain some insight on the issue: LLMs and related technologies increase the ease of combinatorial diversification to an unprecedented scale. You can now generate thousands of versions of a news article within minutes, but many of those articles may contain factual errors, boast fabricated quotes, or simply be boring to read. LLMs threaten to flood the internet with torrents of lackluster prose and become dangerous engines of misinformation.

The missing law also tells us, in broad strokes, what must be done to mitigate this problem: In the face of wild combinatorial experimentation, only through stringent selection can functional information increase. It's up to us—everyday users, AI developers, and policymakers, collectively—to apply the appropriate selection pressures to ensure LLM outputs contain meaningful information before they take flight. The price we pay for accelerated combinatorics is constant vigilance.

Selection and Simplicity

Our world abounds with examples of complexification, but it's also easy to point to examples of systems that grow simpler over time. How does the law of increasing functional information square with sightless fish, phones with fewer and fewer buttons, and artificial evolution experiments that produce shorter and shorter strands of RNA? Are these examples violations of our theory?

Our postulated law of evolution suggests that functional information will increase if the function in question is subject to selection as new configurations are generated. Conversely, it predicts that when a once-vital function is no longer in play, when selection ceases, functional information should no longer increase. Indeed, this is what we observe, especially in the biological realm. In some cases, a once-useful feature will just disappear. Cave-dwelling fish, salamanders, and insects commonly sport atrophied versions of eyes or lack them altogether. There is no selection for eye evolution in the dark.

In other instances, an unused trait can gradually morph, displaying a new function that is subject to new selective pressures. The ancestors of modern-day penguins could fly, but at some point in their evolution, flying conferred no advantage; their wings were no longer needed. Rather, penguins evolved hydrodynamic "wings," an elegant example of exaptation. Now these wing-derived flippers play a critical role in their aquatic lifestyle, and selection will continue to favor the sleekest swimming designs.

Similar changes occur at the level of genes and genomes. Cells that evolve in very stable environments such as deep under the seafloor, or that learn to play specialized chemical roles in an interdependent consortium of cells, often display length selection pressure. Their genomes may become significantly smaller, holding less information, in response to their altered environmental conditions. Such a reduction in genome size may be accompanied by the loss of some functions yet still represent an increase in the functional information of the larger interdependent community.

The same goes for slick touchscreen phones. Modern mobile devices are morphologically simple compared to their Blackberry and flip-phone predecessors, which featured dozens of little buttons crammed into the face of a handheld device. What caused this structural streamlining in the evolution of cell phones? When the relevant function is usability, a touchscreen offers a degree of flexibility that a miniature keypad does not. Touchscreens therefore carry far more functional information than older designs.

In many cases, added complexity affords greater functionality. But when simple things function better, as in the case of cell phones or deep-sea genomes, selection favors simplicity. That's why our proposition is not the law of increasing *complexity* but the law of increasing *functional information*.

Calculating Functional Information

A daunting challenge persists for most evolving systems. Even if the function of interest is clearly in mind, the functional information is virtually impossible to compute. It's easy for us to assert that the functional information of a system is increasing as it evolves, but how do we know that it actually is?

To calculate functional information, you need to know two numbers. The first is the total number of all possible configurations. That's easy to determine when it comes to long strands of RNA letters or lines of computer code, but virtually impossible to estimate in many three-dimensional physical systems, especially those that are constantly abuzz with activity, like collections of cells. And even if you could estimate the total number of configurations, you then need to know what fraction of those configurations achieves a given degree of function. Again, such calculations are impossible for many systems.

Does our inability to calculate functional information in these complex cases invalidate the concept? Based on centuries of applications of other natural laws, such difficulties are by no means fatal. It is often impossible to apply universal natural laws to real-world examples. Newton's universal law of gravity, for example, is fine for calculating the attractive force between two bodies—Earth and the Moon, or the Sun and Halley's Comet, for example. But as soon as a third mass is brought into the picture (the so-called three-body problem), the mathematical calculations typically become intractable.

Consider the asteroid belt, a ring of about a hundred thousand chunks of rock a mile or more in diameter that adopt complex orbits around the Sun. Concentrated in a donut-shaped volume of space between Mars and Jupiter, these rocks adopt complexly changing orbits because of the collective gravitational tugs from many massive objects. There is no conceivable way to calculate the gravitational field produced by thousands of orbiting asteroids. Nevertheless, the concept of gravity and its effects on asteroids is unambiguous. Moreover, we are quite successful at launching space probes to Jupiter, Saturn, and beyond. Exploration of the outer solar system is possible because we can approximate the asteroid belt's gravitational field well enough, even though an exact solution remains elusive. There may be analogous shortcuts for appraising the functional information of complex systems.

But for now, the calculation of functional information is impossible for many evolving systems. We are hard-pressed to enumerate every possible configuration, much less quantify the degree of function for each of those configurations. However, there is at least one natural system where the functional information calculation is tractable: mineral evolution.

Functional Information and Mineral Evolution

The most obvious prediction of the law of increasing functional information is that the functional information of an evolving system subjected to selection must increase. Can we put that prediction to a rigorous test? At first blush that seems a daunting challenge. For most complex evolving systems, we don't have a straightforward way to catalog all possible configurations, much less calculate the functional information of each configuration. Most natural systems, especially living systems, remain too complex for now.

In many ways, mineral evolution is the ideal system to put to the test. The seventy-two chemical elements that form minerals present vast combinatorial possibilities, but not so many that they are beyond enumerating. And, through billions of years of selection, nature has shown us which configurations are functional—which mineral species persist. We also recognize a sequence of stages of mineral evolution—each building on what came before. The unambiguous prediction: Each successive stage of mineral evolution must display an increase in functional information, as measured by the small fraction of all possible chemical combinations that result in a mineral.

Nine sequential stages frame Earth's mineral evolution, starting with the earliest minerals that condensed in the atmospheres of stars, and ending with the mesmerizing mineralogical diversity of Earth today—roughly a third of which are a result of life's activities. The mineralogical literature documents the total number of observed minerals at each stage—those that have been identified and approved as mineral species. But every mineral-forming environment has the potential to create far more combinations of atoms than what we actually find in meteorites and mines. One can calculate the total number of potential minerals based on all possible combinations and permutations of chemical elements at that stage. Consequently, we know both the *possibility* space and the *realized* space of minerals at each of nine progressive stages of mineral evolution. With those two numbers in hand, it's relatively straightforward to calculate the fraction of minerals that have survived over geologic time, and from there the functional information of mineral systems with respect to the function of static persistence. Let's walk through an example slowly.

Consider just one stage—the stardust minerals that formed before our solar system. Three numbers are central to the calculation of potential diversity: (1) the number of different chemical elements essential to minerals at that stage, (2) the maximum number of chemical elements in a mineral formula at that stage, and (3) the largest number of repeated atoms (the highest subscript) in a mineral formula at that stage. Stardust minerals feature a total of sixteen different chemical elements, while a maximum of four different elements is found in olivine [$(Mg,Fe)_2SiO_4$], and the maximum formula subscript is 19 in hibonite ($CaAl_{12}O_{19}$). What is the total number of possible chemical formulas for a system with sixteen elements, up to four elements per formula, and with subscripts from 1 to 19? Standard statistical methods to determine permutations and combinations lead to the answer: 159,184,296 different chemical formulas. By contrast, only twenty-seven different stellar minerals have been identified thus far.

Functional information for this first stage of mineral evolution is easily calculated as the negative log to the base 2 of the fraction 27/159,184,296. The answer: 22.5 bits. Note that this calculation represents the collective functional information of all twenty-seven minerals of this earliest stage of mineral evolution.

Stage 2 minerals—those formed in the early solar nebula—are more complex. As of 2024, fifty-nine Stage 2 minerals had been catalogued, formed from twenty-three different elements. Their chemical formulas have as many as seven combined elements, with hibonite still boasting the largest subscript of 19. The resulting functional information of minerals during Stage 2 is 40.0 bits.

Applying these calculation protocols to the next seven stages reveals 49.3 bits for Stage 3, 57.2 bits for Stage 4, all the way to 138.6 bits for minerals on Earth today. Each stage sees the appearance of more mineral species, but an even greater increase in the potential combinatorial richness. Consequently, stage by stage, a smaller and smaller fraction of all possible chemical formulas appear as stable minerals.

Two important trends lead to this result. On the one hand, through 4.56 billion years of planetary history, the number of mineral species has increased dramatically, from twenty-seven stellar minerals to three hundred meteorite minerals in the earliest stages of the solar system, to more than eight hundred species by the end of Earth's first half-billion years, to an estimated four thousand different minerals formed before the Great Oxidation Event 2.5 billion years ago. Recent estimates suggest that more than nine thousand species occur on Earth today. One might conclude that the *fraction* of all possible mineral species that were generated steadily increased through time. In that case, functional information of mineral systems would have steadily decreased.

What we observe instead is a dramatic *decrease* in this fraction because the rapidly increasing *potential* combinatorial richness of the mineral kingdom significantly outpaces the slower increase in *actualized* mineral diversity. The explosion in the mineralogical configuration space over time is the result of the one-two punch of combinations and permutations: An ever-greater number of different chemical elements leads to the potential for numerous new combinations. At the same time, the complexities of formulas—both in terms of the number of coexisting elements and their largest subscripts—add new permutational possibilities. Consequently, the fraction of all possible chemical formulas that are realized as stable minerals decreases precipitously. And, as predicted, functional information of the evolving mineral kingdom increases.

One of the most remarkable characteristics of many evolving systems is their ability to switch gears and evolve in new ways. Generally, novelty is precipitated by an expansion in the accessible possibility space: Many new configurations become available thanks to the emergence of a critical new building block. In other cases, the environment may change in a way that alters the selection processes.

Mineral evolution provides varied examples, with potential increases in functional information ascribed to at least four factors. (1) Mineral repertoires can expand rapidly with increased diversity and availability of elemental components in the universe, such as new local concentrations of boron, cobalt, or uranium—mineral-forming building blocks that previously had been widely dispersed as trace elements. (2) New minerals emerge because of the ever-expanding pressure, temperature, and composition spaces that trigger mineral formation: for example, during deep burial by subduction and subsequent uplift of a wedge of Earth's crust. (3) Minerals form as the result of the ever-increasing time between crystallization billions of years ago and the present—vast expanses of time during which old minerals gradually transform to other species. (4) Finally, most new minerals arise by modification of the prior generation of minerals, leading to successions of novelty, each of which is only possible once the prior stage occurs.

Bounded Evolution: Studying mineral evolution also reveals an unanticipated result. Functional information increases from each stage to the next, but the maximum value appears to approach a limit—perhaps close to 150 bits. The ninth and most recent stage of mineral evolution features seventy-two mineral-forming elements, with some minerals requiring as many as fifteen different elements and subscripts as large as 135. Even with billions more years of mineral evolution, it's hard to identify many more chemical elements that might make minerals. It's also hard to imagine that mineral chemical formulas would ever become significantly more complex.

Minerals, therefore, constitute a compelling case for a *bounded* evolving system (a confusing, different meaning of the word from the idea of a bounded natural law—a law that only works under certain conditions). No matter what new mineral-forming tricks might be devised, we are now approaching the limit of combinatorial richness.

Nucleosynthetic evolution is similarly bounded. Nature has explored pretty much every possible arrangement of protons and neutrons, resulting in about 340 different stable isotopes. If a system has generated and explored every possible configuration, and if selection has passed judgment on every one of those configurations, then that's it. Lacking the novelty of new types of configurations, for example, by the introduction of a different kind of nuclear particle, a system will become "evolved out." Clearly, there are limits to how evolved certain natural systems can become.

What remains uncertain is whether some evolving systems are unbounded or open-ended—that is, whether they have the potential to continue evolving indefinitely. One plausible hypothesis is that life is open-ended. Another possibility is that life is not truly open-ended, but the combinatorial space of biology is so large that the universe will never have the time to sample all corners of biological possibility. Either way, life's apparent open-endedness may represent a shorthand way to distinguish between living and nonliving evolving systems.

Molecular Biosignatures

The law of increasing functional information has already led directly to one newsworthy prediction and discovery—a new method to detect ancient signs of life on Earth or other worlds. The idea is based on the roles of function and selection in natural systems with lots of carbon-based organic molecules.

Organic molecules occur widely across the galaxy. They constitute an important component of the interstellar clouds of dust and gas that might someday become stars and planets. These molecules also formed and accumulated abundantly in the earliest carbon-rich planetesimals of our solar system. Some of those objects—like the small asteroid Bennu that NASA's ambitious OSIRIS-REx mission visited and sampled in 2020, with precious fragments returned to Earth in 2023—persist today. Other carbon-bearing planetesimals shattered in collisions long ago, some of their fragmentary remains falling to Earth as the diverse class of carbonaceous meteorites. Organic molecules have also concentrated at and near Earth's surface, most profligately in the form of life and its decaying remains.

An enticing opportunity and challenge has been to decipher hidden messages in complex suites of organic molecules. In particular, can we distinguish between an abiotic collection of molecules from a meteorite that has never been alive versus a molecular mixture derived from life—what is known as a molecular biosignature?

The tried-and-true approach, practiced by scores of scientists for much of the past hundred years, has been to scrutinize ancient sedimentary rocks for molecular fossil biomarkers—specific carbon-based molecules that are idiosyncratic to life as we know it. Life's biochemistry produces a wide range of diagnostic chemicals: lipids that form cell membranes; sugars that comprise tree trunks, roots, and leaves; the amino acid building blocks of proteins; DNA and RNA; and hundreds of other specialized molecules used in processes as varied as reproduction, digestion, and photosynthesis. For the adept organic chemist, there are myriad biomarkers for which to search.

A persistent challenge is that time destroys most carbon-based chemicals. They alter and break down, eventually to fragments with just a few constituent atoms and little diagnostic value. Consequently, paleontologists looking for the molecular remains of ancient life focus their efforts on the most pristine, unaltered sedimentary deposits. Even so, for rocks much older than one or two billion years, scant biomarker evidence of any kind remains.

The ancient sedimentary rocks collected by paleontologists present additional challenges. Some potentially misleading organic molecules have the potential to form in the deep crust or mantle by abiotic processes. At the same time, modern fluids bearing traces of the organic chemicals that are ubiquitous in modern environments may contaminate old sedimentary rocks. Teasing out specific biomarkers in the face of molecular decay and false signals is a daunting, and in some cases futile, endeavor.

A related opportunity awaits us beyond Earth: If alien life originated and evolved independently on another world, it may be quite different from life as we know it. Every living thing on Earth descended from a common ancestor and shares a set of fundamental building blocks and common molecular motifs: DNA and RNA as information storage molecules, ATP as energy currency, an alphabet of twenty amino acids, and so on. Some if not all these details may be frozen accidents of how life emerged and evolved early in our planet's history. An alien biosphere, responding to different environmental selection pressures, may have stumbled upon alternate solutions to familiar problems. Thus, for all the marvelous diversity of macroscopic forms that delight our eyes—from lion prides to coastal forests to deep-sea anglerfish—our biosphere collectively represents just a single example of life as it could be in the universe. What may be fundamental to Earthly life is not guaranteed to be universal to life everywhere. Send a DNA sequencer to Enceladus, and it may miss life hiding in plain sight if Enceladean life does not use DNA.

So, what characteristics can we expect all life to share? What can we measure that signifies the presence of life, whether or not it is built the same way we are at the biochemical level? The missing law points to an answer: evidence of selection for function.

A New Approach: In an abiotic system like the ones that produced suites of meteorite organics, the only selection criterion is static persistence. Random chemical processes string together carbon atoms into a bewildering variety of chains, branches, and rings, with a few additional elements—primarily oxygen, nitrogen, and hydrogen—to make the mix even more complicated. Countless billions of different molecules can be generated under such conditions—the only selection criterion being that the molecule doesn't quickly fall apart or react to form some different molecule.

Contrast that scenario with life's mandate to make essential molecules that are selected for their critical biological functions. Survival depends on focusing limited metabolic energy into making more of fewer kinds of molecules compared to an abiotic system.

Two different selection criteria—simple persistence versus biological function—must inevitably lead to fundamentally different *distributions* of molecules. This profound, inherent difference between abiotic and biotic molecular suites suggests a different kind of analytical approach. Rather than search for specific molecular species, we can look at the entire pattern of all the molecules and their frequencies of appearance.

One of the best ways to detect and characterize the organic molecules in a complex mixture is called pyrolysis–gas chromatography–mass spectrometry, or py–GC–MS for short. The name is complicated, but the process is straightforward. Start with a tiny bit of organic-rich material—typically a lustrous black speck no larger than a small grain of sand. Insert that sample into a holder that flash heats the sample to more than one thousand degrees Fahrenheit for several seconds. At this point we've produced a complex mixture of hot gas molecules, many of them remixed and shuffled fragments from our original sample. In other words, the first step of this analytical process creates a molecular mess that isn't even the same as the original messy mixture of your sample. That sounds like a bad thing, right? But it isn't.

The key to this approach is molecular distributions, not specific molecules. We are interested in gathering a vast amount of information about the sample as a whole. With py–GC–MS we get three kinds of information. First, the gaseous molecular fragments are passed through a narrow tube—a gas chromatographic (i.e., GC) column more than one hundred feet long and no wider than a fine hair. The insides of that hollow column are coated with chemicals that slow down the progress of some molecules more than others, thus spreading out the mess of molecular fragments into discrete burps over a period of an hour or so. Each discharge represents one or more molecular fragments that takes a specific time to get through that one-hundred-foot column. These outflows are registered as peaks on a chromatogram—a series of squiggles that record the number of molecules that exit the GC column over time.

A second analytical trick yields even more information. Each of the potentially hundreds of peaks on the chromatogram represents a collection of molecules that is once again fragmented, this time by a powerful electron beam. Then those secondary fragments are sorted by their masses in a mass spectrometer (hence MS). Each of the hundreds of GC fragments produces dozens of diagnostic mass fragments, each with a measured mass and intensity.

The end result is that each sample is characterized by a huge spreadsheet of numbers—a matrix with more than 3,000 increments of time in 3,000 rows, versus about 160 mass increments in 160 columns. In total, each sample subjected to py–GC–MS analysis gives us about a half-million matrix numbers, each number representing the peak intensity for that combination of time and mass. That's a lot of data—far too much to ever look for specific peaks related to specific molecular biomarkers. Instead, we search for patterns. But how does one find a pattern in a half-million data points? The answer is AI.

AI to the Rescue: Identifying a biotic from abiotic signal in a flood of five hundred thousand data points is analogous to the problem of facial recognition. The computer that scans the pixelated version of your face isn't looking for any specific diagnostic mole or wrinkle. Rather, it recognizes patterns among all the data, all at once. Likewise, our method of biosignature detection relies on patterns in all the data from hundreds of samples.

A recent example from archaeology highlights this point. The eruption of Mount Vesuvius that buried the Roman towns of Pompeii and Herculaneum in 79 CE also entombed and baked a collection of papyrus scrolls. For centuries, scholars have tried to no avail to decipher the scrolls' contents from the few remaining irregular ink splotches. Like the molecular bits we want to investigate, these fragments of letters and words seem to contain no useful information—until the power of AI is brought to bear. Artificial intelligence can make sense of subtle patterns that the human eye and brain interprets as random discolorations. To AI, the dots and squiggles become letters and words.

Buoyed by this kind of success, we can collect py–GC–MS data on hundreds of samples, including a variety of modern living plants and animals, some ancient carbon-rich fossils, carbonaceous meteorites, and organic chemicals produced in the lab. Each sample is represented by a spreadsheet with 500,000 numbers; each number is the abundance of a different molecular fragment. We then train an algorithm with data designated "biotic" or "abiotic" and let the machine learn how to distinguish one from another.

The results are astonishing. Trained on a broad range of samples, the best machine learning methods (and there are many versions to try) correctly identify living versus nonliving samples more than 90 percent of the time. What's more, even though our model isn't trained to discriminate ancient versus modern suites of molecules, it clearly distinguishes the degraded molecular suites in fossils from the pristine molecules of modern life.

Subsequent research with hundreds more diverse samples has led to important improvements. Trained with more samples, machine learning can distinguish the molecular patterns of life from nonlife with greater than 95 percent confidence, while teasing out additional attributes of the biotic samples. Even ancient samples far too old to preserve anything like a biomolecule—samples with small molecular fragments analogous to the dots and squiggles on those indecipherable volcano-baked Roman papyri—reveal unambiguous patterns diagnostic of life.

Agnostic Biosignatures: Our AI approach to biosignature detection arose from a specific prediction of the law of increasing functional information. The nature and distribution of life's molecules must be fundamentally different from the abiotic molecular suites of meteorites. This statement says nothing about Earth life. Life on any world in the universe would undoubtedly be a complex evolving system that selects its building blocks for their functions. Life anywhere in the universe must have a molecular distribution different from the untutored mess found in abiotic meteorites. Therefore, we propose that our biosignature is agnostic—it should work for any carbon-based life-form, anywhere in the cosmos.

If we're lucky, in at least one of our lifetimes, samples may be delivered from the surface of Mars to Earth. Some of those samples are likely to contain suites of organic molecules—collections that hold clues to their origins. Applying our method to those samples could lead to at least three outcomes. One possibility is that the molecules are unambiguously remnants of the carbon-rich meteorites that have rained down on the Martian surface for billions of years. That's the conservative scenario that many of us would expect.

But the possibility also exists that Martian organics carry a signature of life. In that case two possibilities, both intriguing in their own way, come to mind. On the one hand, molecular signatures from Mars could look a lot like those from Earth. If so, that finding might lend credence to a long-standing hypothesis that the origin of life occurred on Mars, which is thought to have been warm, wet, and primed for life before Earth. In this model, large impacts on Mars launched chunks of the microbe-bearing Martian crust into space, where at least one living rock found its way to Earth. If so, then we and every other life-form on Earth are Martians.

Alternatively, if life arose on Mars in a process completely independent of Earth's—an origin event with its own idiosyncratic suite of biomolecules—then it's likely that machine learning will recognize Mars life as different. We can be sure that an alien biochemistry will not resemble abiotic meteorite organics. Nor is it likely that a "second genesis" would generate Earthlike suites of molecules.

Such a finding would transform our understanding of life's origin and evolution, from a singular attribute of Earth to a cosmic imperative. A second origin of life would suggest that the law of increasing functional information is more pervasive and powerful than we have realized. It would mean that we live in a cosmos where the continual evolution of protons and neutrons to atoms and isotopes, of atoms and isotopes to minerals and molecules, of minerals and molecules to life, and perhaps from life to learning and science, is an inevitable attribute of our wondrous universe.

Given these supportive findings and the inherent simplicity of the idea that nature reveals two complementary arrows of time—one of increasing entropy and disorder, the other of increasing information and order—we are left with a burning question: Why hasn't anyone said this before?

6

Why Hasn't Anyone Said This Before?

We have proposed that evolving systems are conceptually equivalent in three ways: (1) They are formed from numerous interacting building blocks with vast numbers of possible configurations, (2) there exist processes to generate many of those configurations, and (3) newly generated configurations are subjected to selection.

We claim that a law of increasing functional information describes and explains the behaviors of all manner of evolving systems. Furthermore, the law leads us to numerous intriguing implications and predictions, while providing a convincing framework to codify the behaviors of all such systems.

These ideas are simple, requiring little in the way of specialized vocabulary, no new mathematics, nor any discoveries beyond what has been known for more than half a century. All these statements reflect the well-known behaviors of physical and computational systems that have been studied for decades. Why, then, hasn't such a law been articulated before?

At least four possibilities come to mind, each deserving close attention.

Maybe We Are Wrong

Scientists are skeptics, especially when it comes to bold and largely unproven claims. To propose a missing universal law of information is bold, perhaps audacious to the point of foolishness. Every physicist worth their credentials should be pouncing on our proposals. And our ideas might not be all that difficult to disprove.

Our first vulnerability is the claim of conceptual equivalences among evolving systems. We haven't thought of a counterexample (and we've tried hard), but if there is an example of a natural system that evolves but doesn't conform to our three criteria, then we have created a house of cards and are wrong.

Functional information is another point of vulnerability. It might be possible to demonstrate that some systems evolve in our sense of the term yet decrease in functional information.

Perhaps the greatest vulnerability of our proposed law, with its explicit second arrow of time of increasing information, is that we don't really understand the subtleties of the second law of thermodynamics. Many very smart physicists have concluded that the laws of energy are all you need to explain evolution. These scientists are correct that the second law of thermodynamics is at play in the assembly and operation of every new configuration and in every transfer of energy.

However, we go further. We claim that in any evolving systems where one configuration works better than another, then selection for function must occur—functional information must increase. And that increase in functional information is simultaneous with, but different from, the increase in entropy.

So, where do we stand? We think we're right, but we might be wrong.

It's Already Been Done

We've suggested that what we're proposing—the law of increasing functional information—is new. But what if this idea is basically correct yet merely states what others have already discovered in a different way? A few contenders come immediately to mind.

Decades of research, presented in countless articles and books, have tackled the difficult problem of emergent complexity or emergent systems. For more than half a century, proponents of the field of emergence have pointed to numerous natural and computational systems that display complex behaviors as the result of many interacting particles, or "agents." In some respects, our ideas smell a lot like this storied and sophisticated literature. As in our examples of evolution, emergent systems arise from interactions of lots of components, which under suitable fluxes of energy produce totally new collective phenomena such as sand dunes from sand grains, ant colonies from individual ants, and consciousness from billions of neurons.

The driving force of these emergent systems—the function, if you will—is purported to be dissipation of energy and increase of entropy, as required by the second law of thermodynamics. In a series of influential articles and books, notably by John Holland, Stuart Kauffman, and many of their colleagues who worked at the Santa Fe Institute in New Mexico, scholars have explored the origins and behaviors of varied complex systems, from the fluctuations of the stock market to the origin and evolution of life. Emergence, therefore, is an important variant on the theme of a single cosmic arrow of time.

The ideas expounded by proponents of emergence morph

onto ours in several respects. We both invoke components, configurations, and functions. We both observe change over time, with novel patterns and behaviors as the end result. What is missing, we think, is a clear understanding of the role of information. We see information as playing a central role in cosmic evolution—as fundamental in its own way as energy. Obviously, any evolving system must obey the laws of thermodynamics, just as they do the laws motion and gravity. But nothing in the laws of thermodynamics would lead us to describe and explain, much less predict, that nature had to evolve the conscious brain just to dissipate energy a little faster.

We suggest there must be something deeper—what we see as the actions of selection for functions that enhance persistence and favor advantageous novelty. The invention of local order through increased functional information, not the increase of global entropy and disorder, lies at the heart of our evolving cosmos.

One argument to support our model is the ability of genetic algorithms, those remarkable computer programs that we met in the last chapter, to manipulate and test digital configurations. These algorithms work exclusively by shuffling configurations and testing for function—a computational process that exactly mimics our law of functional information (and Darwinian evolution) but would seem to have minimal connection to the second law of thermodynamics.

Darwin Was Too Convincing

A third possibility that we are wrong is a variant on the claim that our ideas simply repeat what has come before. Many of our concepts echo Charles Darwin and his transformative vision of an evolving biosphere, as argued in his seminal 1859 book, *On the Origin of Species by Means of Natural Selection*. In many respects, our ideas repackage Darwin's exposition of biological evolution by the process of natural selection.

1. Darwin said that individuals of any biological species exhibit variations in their traits; we say evolving systems have the potential to exist in many states.

2. Darwin said that more individuals are born than can survive; we say there must be a process to produce lots and lots of configurations.

3. Darwin said the most fit individuals are the ones most likely to survive and pass on their traits to the next generation; we say evolving systems must have a process that selects for function.

To these three parallel sets of ideas, we add the concept of functional information as a variable inherent in all evolving systems—an idea that surely applies to the biological world that Darwin explored. It was Darwin's contemporary, Czech geneticist Gregor Mendel (1822–1884), who first captured glimpses of the information transfer inherent in all biological evolution. We now know the DNA double-helix molecule carries that genetic information, passing slightly mutated copies from generation to generation.

We've heard two similar kinds of objections to our ideas that refer to Darwin. One opinion is that evolution only applies to life. Mineral "evolution," according to this view, simply isn't evolution. It may involve diversification, there may be congruency, but it isn't evolution. One respected colleague even went so far as to say mineral evolution is "fake science"—a remark perilously close to charges of scientific fraud.

What these critics don't explain is how to handle the complexly intertwined evolution of minerals and atmospheres and life. If the evolution of minerals and life are inextricably connected to each other—if life and rocks coevolve—then how can evolution apply only to living species?

A second, perhaps more subtle and subjective response in support of Darwin's view as the only necessary exposition of evolution, is the conjecture that life is the only *interesting* evolving system. Perhaps it's the only obvious evolving system that has a contingent future. Atoms, isotopes, molecules, and minerals might show characteristics of evolving systems, but if they are deterministic systems—ones where the ultimate distributions of atoms or minerals are preordained—then who cares?

How to respond? Would Newton have said his universal law of gravity only applies to "interesting" systems of three or more objects, in contrast to the rather trivial case of the Earth and Moon? On the contrary, the power of a valid law of nature is that it applies all the time to every relevant system, whether simple or complex.

If there is some fundamental difference between biotic and abiotic evolving systems, then perhaps it relates to the potential for future change. Perhaps only life displays open-ended evolution. If so, then new avenues for research will beckon, but that difference would in no way invalidate our ideas.

A Law of Increasing Function Makes Scientists Uncomfortable

The above three possibilities—that we are merely echoing Darwin, or that our ideas are basically equivalent to those in the field of emergence, or that we are simply wrong—are relatively straightforward to deal with. We've done what all scientists do. First, we put the new idea out there via peer review, in this case published in the *Proceedings of the National Academy of Sciences* in October 2023. Then we wait, giving seminars and writing follow-up papers while letting other scientists consider our proposal. If others don't like what they read, and can show where we've gone astray, then more publications and analysis will follow. Whether we're right, or at the edge of a better idea, or just plain wrong will all be sorted out by the scientific method. We just hope that these ideas are interesting enough to warrant informed conversations and possibly nudge the field forward. Whatever the outcome, we can accept it.

What we cannot so easily deal with are two aspects of the law of increasing functional information that might cause scientists to be wary—more uncomfortable at a philosophical level. First is the concern that in our efforts to calculate functional information, the selection of the function of interest seems to be subjective. We have suggested that functions include static persistence, dynamic persistence, and novelty generation (which may also be a path to persistence). But what is the function of a nitrogenase enzyme outside a single cell, or a single cell removed from its environment? What is the function of a single ant in the absence of an ant colony, or a tree divorced from its forest ecosystem?

The concern is not so much that the calculation of functional information is intractable in such cases; many calculations of quantities related to forces and motions in complex natural systems are similarly elusive. But in those cases, we are dealing with well-defined physical quantities. The mass (in grams) of an object is constant and conserved. The electric charge of a proton or the energy of a chemical bond are well established numbers. The magnitude of an unbalanced force (in newtons) that results in the acceleration of a mass (in grams) is not subjective. But the functional information associated with a cell or an ant or a tree—the exact number of bits involved—is dependent on the context. The way you ask the question may affect the answer. That is a problem for most scientists, including us.

Our tentative answer, if not a satisfying solution, is that selection for function is always context dependent. Consider an everyday example. Think about the coffee cup sitting beside you as you read this book. That cup is made of perhaps 10 to the 25th power atoms—building blocks that can adopt truly astronomical numbers of different configurations. Only a tiny fraction of those configurations would be able to hold coffee with a convenient handle; therefore, the functional information of those atoms as a coffee cup is relatively high (though even a rough estimate would require a lot of assumptions).

Alternatively, if you're outside and there's a breeze blowing, then you might be using your coffee cup as a paperweight. A coffee cup works reasonably well as a paperweight, though it's by no means the optimal design using those 10^{25} atoms. Consequently, the functional information of your coffee cup as a paperweight is much smaller (i.e., fewer bits of information) than as a coffee cup. And with respect to other functions—as a screwdriver, for example—the cup has zero functional information. In other words, the functional information of an object is contextual. Unlike mass or charge or force, which are independent of context, functional information requires the observer to choose which function is of interest. The task is subjective. And that is, indeed, a real problem for scientists.

The same kind of contextual considerations apply to any complex evolving system. One season an evolving plant species might be selecting for resistance to drought, the next season to flooding; one year that plant might have to deal with unusual heat, the next year freezing cold. Contexts change as environments change—a fact mirrored in the complex and seemingly unpredictable way things evolve. In a complex evolving system, the function being selected may change with context.

Is this a fatal flaw in our model? Is it possible for a sound scientific hypothesis to incorporate subjective analyses of context and function? Such subjectivity would seem anathema to scientists because "selection" and "function" hint at the possibility of purpose and value in specific scientific contexts. So, maybe we are wrong.

On the other hand, maybe, in the case of complex evolving systems, some degree of subjectivity is essential to understanding. As contexts change, selection for function must also change. As we've said before, complex systems are complex. As objective scientists, our task is not to describe nature in a way that is convenient or that conforms to some philosophical preconceptions of the ways we think things should work. Our responsibility is to describe and explain nature in all its complexity, as it "really" is.

A much deeper philosophical challenge arises from the sense that our proposed law suggests not just change over time but also hints at something akin to quantifiable "progress." Functional information increases with time. Systems improve, as definitively measured by an increase in function. Our view is of a world that keeps doing more and keeps doing it differently. Can anyone honestly argue that the universe has not gotten a lot more interesting over its 13.8-billion-year lifetime? Our law codifies that trend, while predicting that it will continue to do so until such a point when (if?) the universe ceases to evolve.

Don't worry—that won't happen for a very long time. And in the meantime, the law of increasing functional information presents us with profound implications and opportunities.

7

Implications: What Does the Missing Law Mean for You?

Implicit in the law of increasing functional information are many consequences regarding the nature and behavior of the wide range of evolving systems. One of the most obvious implications is that biological evolution, which Darwin expounded so lucidly in the nineteenth century, is a special case of a more general phenomenon. We see now that a broader concept of evolution—one based on the processes of combinatorial genesis and selection for persistence-enhancing and novelty-generating functions—applies not only to the living world but also to all kinds of physical and chemical systems, from mineral evolution to the evolution of computer code.

The law of increasing functional information not only changes the way we contextualize Darwin's theory. It also has profound implications for the way we understand time, our place in the universe, the origin of life, and purpose.

Nature Reveals a Second Arrow of Time

An obvious aspect of evolving systems is that they have a direction in time—what physicists call temporal asymmetry. Evolving systems begin modestly, with low diversity and relatively simple forms. The starting point for the evolution of the elements and isotopes was a mix of hydrogen and helium. Organic chemistry—the complex chemistry of the element carbon—began with such simple two- and three-atom molecules as carbon monoxide and water. And the remarkable diversity of minerals on Earth commenced with the formation of a couple of dozen ur-minerals. Populated with these initial building blocks, each system displays increases in diversity, distribution, and/or patterned behavior. These examples and many more underscore the point: The missing law of evolving systems has a built-in arrow of time.

The temporal asymmetry of the law of increasing functional information is fundamentally different from the second law of thermodynamics—the one that posits an increase in disorder with time. Nevertheless, we suggest that information behaves in an analogous way. In fact, we see a possible dramatic parallel between the laws of energy and the laws of information.

The German physicist Rudolf Clausius (1822–1888) is often credited with the first formal statement of the second law of thermodynamics, that entropy tends to increase. In fact, he stated succinctly:

The energy of the universe is constant.

The entropy of the universe is increasing.

Might there be a parallel kind of statement related to information? Indeed, might there be more than one natural law of information? We speculate that:

The Kolmogorov complexity of the universe is constant.

The functional information of the universe is increasing.

This proposition, though untested, makes sense to us. Kolmogorov complexity is the total amount of information required to completely define a system. If that system is the universe, and if the total amount of mass and energy is fixed, then the cosmos may adopt many different configurations, each requiring the same amount of information to describe the system. If our conjecture is valid and the number of bits required to describe the universe is constant, then there can be no arrow of time in the context of Kolmogorov complexity.

Functional information, by contrast, is context and time dependent. Whatever function we use to characterize the universe—clumpiness, diversity of objects, persistence, self-awareness—as long as selection is in play then the associated functional information must be increasing.

The law of functional information describes an increase in order. It's a metaphorical counterbalance to the second law of thermodynamics. If poetically inclined, one could think of these two laws as the yin and yang of time. The second law describes the destination. The law of functional information describes the journey. Working together, both laws help explain and predict the behavior of phenomena in our everyday lives.

New Views on Life's Origins

The question "How does life begin?" has perplexed scientists and natural philosophers for eons. At root, the origin of life is a complex problem in chemistry. Varied chemicals must mix and interact until some felicitous combination learns the remarkable trick of making copies of itself. Once that happens—once a chemical system becomes self-replicating—then it is well on the pathway to Darwinian evolution by natural selection.

What does the law of increasing functional information add to this story? It suggests that we can understand the lead-up to life's onset—the period before Darwinian evolution set in—through our expanded concept of evolution by selection for function. In this context, life's origin was not the one and only transition to an evolving world. Rather, the origin of life was but one of many steps in the evolution of the cosmos.

An almost universal assumption in the origins-of-life research game is that carbon chemistry is the only logical foundation. Of all the hundred or so known elements, only carbon has the astonishing versatility to build molecules of virtually any size and shape, with almost any imaginable chemical behaviors. This versatility stems from the combinatorial richness of carbon-based structures. Carbon can bond to itself as well as to most of the other elements of the periodic table; it can form chains, rings, branching structures, and any imaginable combination of those motifs. Carbon-based molecules were abundant on early Earth, and they constantly experimented with new forms and combinations.

The configuration space of possible organic compounds alone is vast, but adding the inorganic world of minerals to the mix greatly enhances combinatorial richness. Minerals also give organics a lift when it comes to generating new possibilities. Many of the clay minerals present on early Earth contained iron, nickel, and other metal elements that are famous for their catalytic, molecule-enhancing properties. Moreover, clay minerals' sheetlike atomic layers provide surfaces where organic molecules can congregate and interact in new ways with greater frequency. Hence, chemical transformations that are sluggish in the absence of clay minerals often proceed lickety-split when clays are around. Metal-bearing clay minerals could have been the key to driving new organic molecules into existence.

Origin-of-life scenarios can be complicated and perplexing, filled with long names of participating organic molecules. But when you look past the details, you find underlying similarities across all origin-of-life scenarios—three fundamentals we all agree upon:

1. For life to begin, the environment needed to supply *a multitude of diverse components*. Perhaps atmospheric chemistry and impactors from space delivered a dizzying array of organic compounds to shallow pools on the surface of the Earth, while reactions between seawater and Earth's early crust produced a panoply of organic molecules within the plumbing systems of hydrothermal vents.

2. For molecules to complexify, the environment needed to feature *processes that pushed starting blocks into myriad new configurations*. Ultraviolet light might do the trick. So might the vigorous mixing of fluids in hydrothermal

vents or the shuffling of atoms and molecules on mineral surfaces.

3. For life to evolve, the environment needed to *select certain molecular configurations in preference over others.* Suites of molecules that replicate with more fidelity and chemical cycles that are more resilient would have outcompeted less successful constructs.

These three criteria underlie all evolving systems: a diversity of components, ways to generate new configurations of components, and selection for function. Although there is still significant disagreement regarding the most promising early Earth environment for the origin of life, the most viable abiogenesis theories are united by their adherence to the principles above. The origin of life was not the beginning of evolution; it was just another step along a continuum of change.

Chance and Necessity

A strange dichotomy shadows evolution. On the one hand, evolving systems rely on the random sampling of vast numbers of possible configurations out of a much larger population. In most systems of interest, the number of possibilities is vastly greater than what could be generated by all particles over the lifetime of the universe.

Consider a modest-sized enzyme—a chain of 100 amino acids. Each of the 100 positions can feature any of 20 different biological amino acids, each with a different size, shape, and chemical character. That means there are 20 to the 100th power possible chain-like configurations. That number, 20^{100} or roughly 1 followed by 130 zeros, is beyond astronomical. It is a trillion-trillion-trillion-trillion times greater than the estimated number of electrons in the entire universe. Even if a trillion galaxies, each with a trillion planets, manufactured new enzymes at a prodigious rate, only a tiny fraction of all possible 100 amino acid configurations could be sampled in the lifetime of the universe. At least on the surface, chance appears to play an outsized role in such combinatorically rich systems.

On the other hand, the outcomes of some evolving system appear to be predetermined—deterministic. Despite the combinatoric richness of proton-neutron configurations, we find that the same modest numbers of elements and isotopes always appear in stars. We are equally confident that certain groups of minerals—common rock-forming pyroxenes, micas, feldspars, and garnets, for example—will always emerge on terrestrial planets. In the molecular realm, experiments in aptamer evolution consistently zero in on the same few classes of solutions—the same small clusters of chosen configurations out of more than 10 to the 60th power of possibilities. Similarly, genetic algorithms converge, despite the almost infinite variety of possibilities. In these and many other examples, the final answer seems to represent an inevitable, deterministic solution.

Vigorous scientific debates have explored this question of chance versus necessity, especially in the context of life. The noted American paleontologist and evolutionary biologist Stephen Jay Gould (1941–2002) argued strongly in his 1989 book, *Wonderful Life*, that life is replete with contingency. Play the tape over again, Gould posited, and an entirely different menagerie of beasts with completely different body plans would populate Earth. Gould's convictions were countered by British paleontologist Simon Conway Morris (b. 1951), who saw convergence to a few highly functional designs as the rule. Many of us see both chance and necessity playing key roles: Life is flamboyantly varied, yet many features—eyes, wings, fins, and legs, for example—evolve over and over again because they work.

We are left to wonder if determinism is a predictable aspect of evolving systems, or if one just has to wait and be surprised, like by the design of NASA's quirky ST5 space antenna. And, if we understood complex evolving systems such as life more deeply, would some degree of determinism emerge?

The answer may lie in the exploration of function landscapes. Imagine representing all possible configurations of a system on a large surface—a plane of possibilities. In reality, such an *X-Y* plane is a poor substitute for the multiple dimensions of complex systems, but you can at least imagine reducing all those dimensions to a kind of landscape. Every point on that *X-Y* plane represents a configuration, and every one of those configurations is associated with its own value of functional information as plotted as a height in the *Z* direction—the elevation of the topography, so to speak. In this imaginary landscape, most points have zero function, so they are represented by a flat expanse that may cover most of the area in a featureless plane. However, some points have function and so can be represented as raised areas of the landscape—hills and peaks of function. And that is where chance versus determinism comes in.

For some evolutionary challenges there is one and only one solution. If you mix the elements silicon, zirconium, and oxygen together in all sorts of ratios, for example, you'll get the mineral zircon almost every time. That means that on the very limited function landscape that represents different combinations of those three elements, there is only one primary sharp peak—the one corresponding to $ZrSiO_4$. That is determinism. One solution emerges every time. In other instances, as in the case of mixing protons and neutrons, there may be dozens or hundreds of sharp peaks dotting the landscape, each representing an exact integral combination of nuclear particles—each one of the 340 stable isotopes. That, too, is determinism.

Not all landscapes are so spiky. You can imagine function landscapes with gently rolling hills or several craggy mountains or ridgelike connections among several high points. We suspect that function landscapes representing organic molecules in carbon-rich meteorites, with literally millions of different configurations having similar long-term stabilities, might look something like that. Alternatively, you might have two widely separated steep-sided peaks, with a series of lower hills and high points dispersed about—perhaps applicable to Jack Szostak's aptamer experiments. In those cases, there exists a real element of chance in the final configuration. To be sure, you will inevitably wind up at some high point, but which one may be a matter of chance—perhaps dependent on where the first weakly functional arrangement appeared on the *X-Y* plane and how it gradually increased in height—an increase in functional information, as the new law demands.

Function landscapes may also change over time. As environmental conditions evolve, what may once have been the Everest of functionality can erode into a small hill or even a valley. For systems whose function landscapes change rapidly, the ability to jump around in search of new heights is beneficial. In other words, novelty generation becomes a powerful selective force. Novelty generation does not play an outsized role in deterministic systems, like atoms and minerals, whose function landscapes are static and spiky. But when it comes to living systems, whose function landscapes are complex, chaotic, and ever-changing, creativity is key to persistence.

Why does biological evolution continually produce new richness, rather than solve for the most optimal solution and call it a day? The answer is that Earth is complexly dynamic, and life itself compounds that complexity. The function landscape that life must navigate is not only rugged, with many precipices and false summits, but also *changing* in time. In a variegated environment, there's always something new to learn. It pays to be able to predict changes, to find new resources, and to develop new ways of being. Over generational time, Darwinian evolution produces recipes of success written in genetic code. But it's also a slow and wasteful endeavor. So, life evolved faster ways to sense and transmit information about its environment. Smell, sight, speech, schooling, silicon processing—each of these functions serves life's inexorable quest to digest and transmit information so that it can persist in a dynamic, changeable world.

Life on Earth has burst into a dizzying array of being and becoming. But one thread ties you and every living thing around you back to our humble beginnings: information. You carry within you the ultimate molecular memory that was forged into existence at the origin of life. A memory of *persistence*—of staying the course on a rocky sea of surprises and fluctuations. A memory of *replication*—of gathering carbon-rich components to build oneself over and over and over again. A memory of *novelty*—of letting the winds of chance take you to places you've never dreamed.

Cherish the Present

Macroscopic natural laws have limits beyond which they don't work. Newton's laws of motion only apply to objects traveling much slower than the speed of light. The universal law of gravity only applies to objects with masses much smaller than black holes. Similarly, the law of increasing functional information only operates if certain criteria apply. Such examples are referred to as bounded laws (not to be confused with bounded evolution, discussed in chapter 5, which describes a system with limited evolutionary potential).

The law of increasing functional information is a bounded law because at least two circumstances can cause a system to cease to evolve. On the one hand, a system cannot evolve if no new configurations are being generated. In the case of minerals, if a planet completely freezes, then mineral evolution is literally stopped cold.

Alternatively, a system cannot evolve if there are no selective pressures at play. A computer can spit out billions of random strings of letters or numbers, but without some criteria to select one configuration in favor of another, you're left with nothing but a lot of random strings of letters or numbers. Likewise, organic molecules are produced with prodigious variety in the parent bodies of carbon-rich meteorites, but in space there are no selective pressures to favor higher levels of organization in these molecular systems. Persistence is the only criterion for survival.

Leading cosmological models predict that in some 10^{100} years, give or take a few zeros, our ever-expanding universe will approach a so-called heat death, cooling indefinitely into the depths of time. At this point, the universe will be near its maximum-entropy state. All conceivable reserves of useful energy will be spent, curtailing combinatorial exploration. In the end, little will remain except for supermassive black holes slowly evaporating into a dark, uniform bath of thermal radiation. If this picture comes to pass, the law of increasing functional information will no longer apply. Every pocket of the universe will be outside the bounds of the law.

Of course, you needn't worry about that far-off future. Countless complex systems will emerge, evolve, and be extinguished between now and then, and some astrophysicists have suggested that the universe can produce an infinite number of computations before heat death.

Nevertheless, thinking about our cosmological bookends—the universe's scalding birth, too chaotic for order to emerge, and the universe's frigid fate, too inert and uniform to produce complexity—gifts us a profound perspective on the present: Cherish today, for we truly belong to the interesting part of cosmic history, a middle age in which the law of increasing functional information manifests in a most dazzling array of phenomena, including ourselves.

Does Function Imply Purpose?

As the overall entropy of the universe continues to increase, pockets of the universe—from minerals to computer code—can also increase in functional information over time. If what we've observed so far continues to hold, then one thing we may have to jettison is a view of a universe devoid of function and purpose.

Invoking selection for function as the driving force behind

complex evolving systems in all living and nonliving realms raises the potentially uncomfortable question of teleology—the possibly misguided tendency to explain phenomena in terms of the perceived purpose they serve, rather than the natural cause by which they arise. If natural laws demand that our universe transforms over and over again, from atoms to stars to planets to life to consciousness and the quest to understand all that came before, then is there any evidence that nature has an underlying purpose?

Science has no qualms about describing and explaining the increase in disorder that arises from the second law of thermodynamics—the singular law that explains how a hot cup of coffee cools and steam engines dissipate heat. The increase in disorder is a palpable drumbeat of our existence—an impartial, ubiquitous reminder of death and decay. As objective scientists, we accept without emotion that such a law is ever present and that its consequences are irreversible.

Is it so difficult, then, to imagine that the lawful universe also incorporates a process of increasing information and local order? Or, to put it another way, must we conclude that the increase in information and order that we see all around us is nothing more than a quirky spinoff of increasing entropy?

Science is supposed to be objective and impartial, with reproducible observations and experiments that lack any hint of bias. Consequently, many scientists are uneasy at the invocation of seemingly subjective claims of function and selection as driving forces in the natural world, much less that the universe displays some sense of progress. But we are in murky waters here. The most basic definition of "function" as a noun is the purpose for which something is designed or exists. Yet for many scientists, "purpose" is a word that must be avoided in serious scientific discourse.

Does function imply purpose? We offer two answers, one to reassure our scientific colleagues, the other to inspire deeper conversations about the contrasts among what scientists want the universe to be, versus the mindset of theologically inspired thinkers, versus what the cosmos might actually be.

Function Does Not Imply Purpose: The macroscopic laws of nature describe, explain, and in many cases predict the behaviors of matter, energy, forces, and motion at everyday scales of space and time. Each law applies to a part of the cosmos, while taken together they are intended to provide a complete explanation and description of what we experience in the physical world. In this context, none of the laws implies purpose—they simply codify what is.

Take gravity, for example. The "purpose" of gravity is not to make stars and planets, even though the formation of a star or a planet is a consequence of gravitational forces. Likewise, the purpose of the electromagnetic force is not the formation of chemical bonds, much less the electric motor or electric generator, even though chemical bonds, motors, and generators are fully described and explained by the laws of electromagnetism.

In that spirit, if the law of increasing functional information is correct, then it provides a description and explanation of how evolving systems in the universe emerge and how they work. The law recognizes that selection among different configurations is important. It says that some configurations are quantitatively better than others under certain selective pressures. But the law of increasing functional information says absolutely nothing about the purpose of evolving systems. Such systems are, like planets forged by gravity or chemical bonds formed by electromagnetic forces, just a consequence of lawful universal processes.

We conclude that the laws of nature, individually and collectively, need be nothing more than useful tools to describe, explain, and predict the past character, the present state, and the evolving future of the cosmos. In that case, a new law that clearly articulates the exact process by which many local systems display increased order and information is an essential part of the story.

Function Does Not Preclude Purpose: Matters of philosophy and faith evolve. There was a time, not so long ago, when natural philosophers saw purpose in every phenomenon and law of nature. Newton and his contemporaries assumed that every chain of causes must originate with the "self-created being" (that is, God). In this cosmic view, dating back to ancient Greek philosophers, the "first cause" lies at the heart of every aspect of nature, including Newton's laws of motion and gravity.

As recently as the 1840s, James Prescott Joule described his experiments on the conservation of energy and the first law of thermodynamics in teleological terms:

> *Nothing is destroyed, nothing is ever lost, but the entire machinery, complicated as it is, works smoothly and harmoniously. . . . Everything may appear complicated in the apparent confusion and intricacy of an almost endless variety of causes, effects, conversions, and arrangements, yet is the most perfect regularity preserved—the whole being governed by the sovereign will of God.*

Few scientists would write like that today. Something in the way science is done, or the way scientists think, or the polarization of scientific and theological epistemologies, has all but expunged the word "God" from scientific discourse. Perhaps this change reflects an increasing demand for rigorously reproducible, independently verifiable experimental and theoretical methods. Perhaps an atheistic perspective grew out of the existential angst of a generation of twentieth-century physicists who had to grapple with the puzzles of quantum uncertainty while contemplating their central roles in the threat of nuclear annihilation. Whatever the reasons, discussions of God are not part of the modern scientific conversation.

Whatever your philosophy, the laws of nature can neither prove nor disprove the existence of a Creator. They simply describe and explain the behavior of systems of matter, energy, forces, and motions.

If you are agnostic or atheistic, then nothing in the canon of natural laws need trouble you. The laws of nature can be viewed as nothing more than an instruction manual for the workings of the physical world. If there is no God, then you should understand the laws of nature, because they describe absolutely everything you can and will experience.

On the other hand, if you believe devoutly in a Creator, then nothing in the canon of natural laws need trouble you. The laws of nature can be viewed as a revealing guide to how God is manifest in an evolving, physical world. If there is a Creator, then you should understand the laws of nature, because they provide the best tangible window into the mind of God.

So, let us set the question of God aside for now. Is it possible to recognize and understand purposefulness in the universe, decoupled from the existence of a divine Creator? That is, can meaning and value be a part of our scientific description of the natural world? The law of increasing functional information suggests that the answer might be yes—that function and purpose could be emergent properties.

Reality can be described using the language of science in many different ways. In physics, for example, we use different equations for quantum (tiny), classical (everyday life), and relativistic (enormous) phenomena. Biology, too, appeals to different concepts depending on whether we are concerned with protein folding (subcellular), embryonic development (multicellular), or ecological succession (multispecies). Each of these scales comes with its own set of bounded laws that apply brilliantly in that—but only that—regime. In other words, science offers us a pluralistic toolkit for describing the natural world.

When it comes to explaining everyday phenomena, it is almost always *better* to use macroscopic laws (those listed in chapter 1) than the laws of, say, particle physics or quantum mechanics. We can predict a river's flow without tracing every individual molecule of water. We know a freshly baked apple pie will cool down without watching each of its microscopic particles gradually decrease in velocity. It would be folly to describe the outbreak of a pandemic, or even why a square peg does not fit into a round hole, using quantum mechanics. Why? It turns out that there are emergent macroscopic regularities that we can observe, characterize, and describe in a lawful manner without having to appeal to any underlying microscopic explanation. A fluid's *viscosity* or a pie's *temperature* or a person *being infected with a virus* are not concepts that appear at the most reductionistic level of physics. They are properties that only emerge when many, many particles get together to form familiar macroscopic entities. Yet they are genuine features of reality.

While everything might reduce to the quantum level in the end, the emergent layers of reality offer much more economical—and much more predictively powerful—kinds of scientific descriptions. So, we *could* attempt to describe the evolution of minerals, life, language, and technology using the principles of quantum mechanics. (No one is stopping us from trying!) If there weren't genuine regularities at higher levels of reality, then reductionism to the quantum realm might be our only hope. Thankfully, as Darwin showed for the case of life, there are much simpler descriptions of evolution ready to be discovered.

We posit a law of evolution built upon the ideas of function and selection—concepts that don't appear in basic physics. Nonetheless, like viscosity and temperature, function and selection may be authentic features of the natural world. In proposing a new law of nature, we are not suggesting that anything about the underlying laws of physics requires revising. Rather, we are saying that evolving systems constitute a cross-cutting domain whose behavior can be captured by a unifying statement—the law of increasing functional information—that involves contextual properties absent from previously articulated laws.

Let us return to the problem of purpose in nature. As we have argued over the course of this book, understanding complex evolving systems requires us to understand how functions arise that contribute to persistence and novelty. If the phenomenon of function is genuine and scientific, could purpose be, too?

Recall the nitrogenase enzyme—the molecular machine that turns nitrogen gas into biologically essential ammonia. How do we best describe the reason why nitrogenase enzymes exist? In the language of the missing law, they serve a persistence-promoting function. In general, the *purpose* of a biomolecule is something that emerges in the contextual arrangement of matter that we call a living organism. The *value* of a symbiotic relationship among species might emerge in the context of an ecosystem seeking to achieve dynamic persistence. The *meaning* of a geologic process could emerge from its role in maintaining certain global conditions—or its ability to open novel niches.

Science has long presented us with a mechanistic world, rather than a world of purpose. But why can't it be both? Why can't the natural, lawful, generative forces of the universe also produce a cosmos alive with meaning and contextual, relational value? These are aspects of the universe that we all know, feel, and experience—but which science is only beginning to comprehend.

What You Know to Be True: We have no special insights that might persuade you whether or not to believe in a Creator. What we can emphasize is what you already know deep down. We all experience two arrows of time, every moment of every day of our lives.

The second law of thermodynamics codifies our experiences with increasing disorder—scuffs and scrapes, bumps and bruises, debilitating sickness and broken bones, deteriorating eyesight and declining hearing, aching backs and weakening limbs, dementia and death. This dark, depressing arrow of time is all too present in our lives.

At the same time, even as entropy incessantly increases all around you, Darwin reminds us that "endless forms most wonderful and most beautiful have been, and are being, evolved." Look around you. The consequences of evolution are everywhere, all the time. The universe started almost fourteen billion years ago with the Big Bang—a time when there were no atoms or molecules, no planets or stars, no people nor any books that explore the Big Bang. Then came stars and the production of more than 100 elements and 300 stable isotopes. Planets inevitably followed and with them the evolution of oceans, atmospheres, thousands of minerals, and ultimately life.

Single cells evolved new chemical tricks, while clumping into ever more complex plants and animals. Life in the sea expanded to life on land and in the air. Consciousness evolved, along with language, music, science, and art. And now we watch, uncertain, as artificial intelligence evolves in new and unpredictable ways.

The law of increasing functional information attempts to expand and correct the scientific description of nature's changing face—changes that have long been explained exclusively in the context of increasing disorder. Evolution is not, at root, a process of increasing disorder, even though increasing disorder at the global scale is present in any evolutionary process. Rather, the hallmark of evolution is increasing local order, as measured by an increase in an evolving system's functional information. This increase in local order and information is, to us, an undeniable second arrow of cosmic time.

And that is why we are proposing a trio of bold changes to the existing scientific paradigm, the three central points of this book:

1. *There exists a law of nature that describes the generation of order in a world of decay.* We have proposed a law of increasing functional information, a metaphorical counterbalance to the second law of thermodynamics. The latter describes the destination; the former describes the journey. In isolation, each law explains only a fraction of our marvelous world. But together, they revolutionize the way we understand the creativity of the cosmos.

2. *Evolutionary theory should expand beyond, while being inclusive of, biological evolution.* Life is the quintessential complex evolving system, but biological evolution

represents just one small slice of a much wider spectrum of evolution. Nonliving systems evolve, too, as typified by the evolution of atoms and minerals, and so do symbolic and technological systems, from languages to AI. We propose that all evolving systems share certain conceptual equivalences that unify these disparate phenomena under a common conceptual framework.

3. *Information is as vital a parameter in the natural world as mass, charge, or energy.* Science measures properties of natural systems, and through these measurements derives principles that describe their behavior. The scientific canon includes laws that consider physical attributes like inertial mass, electric charge, and kinetic energy. So far, a well-accepted law of information has yet to burst upon the scene. But we propose that a scientific description of the macroscopic world is incomplete without one. Information might even be the secret to quantifying and codifying evolution.

EPILOGUE

Choices

We cannot avoid an increase in entropy in our lives, but we can choose to act in ways that minimize some of its worst consequences. Drive safely. Look both ways before crossing the street. Floss.

We can also actively, consciously choose to increase the functional information of our lives. All of us are faced with difficult choices, many of them consequential. Like evolving minerals or cells, our world of combinatorial possibilities is vast. We have the power to sort through those possibilities, to select pathways that are most likely to benefit others and ourselves, while rejecting words and deeds that are likely to increase the pain and unhappiness of others.

If you've ever woken up with a sense of purpose—whether it's a desire to put in an honest day's work, plan a surprise birthday party, or just find the nearest toilet—you know that goals are a very real part of life. Many of these goals come from a desire to persist; we want to be functional parts of society, valued members of our friend groups, and healthy human beings. Such feelings are natural for a species that evolved in small bands of hunter-gatherers. But today, humanity is a *planetary* phenomenon. We have terraformed our world with engines and electric grids, forestry and fertilizer, concrete and carbon emissions. Our goals, as a consequence, must become more thoughtful and expansive.

Integrated, as we now are, into Earth's biogeochemical systems, many of the relevant selection pressures on humanity's persistence exist at the planetary scale. Our species has transitioned from a function landscape of tribal competition to a function landscape defined by the health of the global systems in which we are a part. We must learn to function in ways that positively impact not only our local neighbors and communities but also the global environmental processes upon which our collective well-being relies. In this newfound context, raising one's functional information means learning to feel a planetary-scale sense of purpose, too.

Unlike the second law of thermodynamics, time's arrow of increasing information gives all of us a degree of agency—the chance to select our favored path in life. In this way, the law of increasing functional information could be seen as a law of hope. Yes, we will, each of us, be battered by the incessant drumbeat of entropy. We must all eventually succumb to its ultimate demand. No one gets out of here alive.

At the same time, during our brief moment in the sun, we can choose to treat those around us with thought and care. We can choose to build, to teach, to explore, to create, to love. All of us are born with the capacity to spark joy—to bring new knowledge, new stories, new meaning into the cosmos. In a universe of inevitable disorder yet increasing information, we can choose to live lives that enrich the order of our world.

You are the latest chapter in the long story of evolution that extends all the way back to the Big Bang. Despite the constant drumbeat of entropy, you have agency. You can imagine the possibilities of your life, select your path, and strive to add to the ever more spellbinding and magnificent unfolding of our evolving universe.

Acknowledgments

This book would not have been possible without the collaboration of our friends and colleagues in the "Missing Law Group," including D. J. Arends, Stuart Bartlett, Jim Cleaves, Carol Cleland, Heather Demarest, Jonathan Lunine, and Anirudh Prabhu. Our years of Zoom and in-person discussions and debates, notably during the time of Covid lockdowns in the early 2020s, helped to define and refine many of the ideas presented here.

The development of these ideas benefited immeasurably from extensive conversations, feedback, and other exchanges with numerous colleagues, including Chris Adami, Danica Adams, Gunnar Babcock, Bruce Alberts, Saleem Ali, Blaise Agüera y Arcas, Kevin Arnold, Philip Ball, Mark Bedau, Steve Benner, Lee Billings, Elisa Biondi, Jonathan Blake, James Brent, Joy Buongiorno, Katy Cain, Sean Carroll, Francesca Cary, Eric Chaisson, Alexandre Champagne-Ruel, Maddy Christensen, Milan Ćirković, George Cody, Elise Cutts, David Deamer, Ahmed Eleish, Adam Frank, Gourab Ghoshal, Faye Flam, Nils Gilman, Donato Giovannelli, Stephen Godfrey, Amit Goldenberg, Andrea Goltz, Michael Gorman, Julien Gough, David Grinspoon, James Gross, Dmitri Gunn, Doug Hamilton, Grethe Hystad, Eric Isaacs, John Jaszczak, Kate Jeffrey, Sarah Stewart Johnson, Stephen Johnston, Stu Kauffman, Shi En Kim, Corey Knox, Artemy Kolchinsky, Richard LaBrie, Émilie Laflèche, Dante Lauretta, Dominic Legge, Rhys Lindmark, Ambrose Little, Karen Lloyd, Cynthia Lunine, Sarah Marzen,

Margaret McFall-Ngai, Daniel McShea, Natasha Metzler, Shaunna Morrison, Mahin Mursalin, Vikas Nanda, Alexandra Ostroverkhova, Howard Page, Medha Prakash, Anjali Piette, Gary Puckrein, Roberta Raffaeta, Anselm Ramelow, Paul Rimmer, Dyna Rochmyaningsih, Sarah Rugheimer, Mike Russell, Alexa Sadier, Caleb Scharf, Corday Selden, Jaylen Shawcross, Neil Shubin, Eric Smith, Kayla Smith, Damian Sowinski, Jack Szostak, Frederic Thomas, Paul Voosen, Melvin Vopson, Mike Walter, Richard Watson, Claire Webb, Tamsen Webster, Jasmina Weimann, Loren Williams, Stephen Wolfram, David Wolpert, Philip Woods, and Yuk Yung.

Many of the interactions that fueled this work would not have been possible without generous funding from the Carnegie Institution for Science, the John F. Templeton Foundation, the Alfred P. Sloan Foundation, the NASA Hubble Fellowship Program, and the NASA Topical Workshops, Symposiums, and Conferences Program.

Early versions of the manuscript were significantly improved by perceptive reviews from writers Margaret Hindle Hazen and Elizabeth Hazen.

We thank the enthusiastic and professional team at W. W. Norton. Our editor, Jessica Yao, played a major role in shaping the book, both through her high-level strategy regarding content and style, and her thoughtful, detailed editorial advice throughout the writing process. Copyeditor Charlotte Kelchner provided an invaluable polish to the manuscript. Gwen Cullen, Susan Sanfrey, Derek Thornton, and Yumiko Gonzales Rios effortlessly transitioned the book from manuscript to print. Siena Ballotta German and Steve Colca guided us through a dynamic and effective promotional campaign. We are also indebted to Eric Lupfer, our tireless literary agent and thoughtful, creative partner at United Talent Agency.

Notes

1. THE LAWS OF NATURE

1 **The macroscopic natural laws:** Robert M. Hazen and James Trefil, *Science Matters: Achieving Scientific Literacy*, 2nd ed. (Doubleday, 2009). For a more detailed review of the history and content of the macroscopic laws of nature, see John L. Heilbron, *The Oxford Guide to the History of Physics and Astronomy* (Oxford University Press, 2005).

5 **Religious precepts:** Mary Midgley, "Heaven and Earth: An Awkward History," *Philosophy Now* 34 (2001): 18–21.

5 **Many scholars:** Carlo Natali, *Aristotle: His Life and School* (Princeton University Press, 2013).

6 **Galileo Galilei:** David Whitehouse, *Renaissance Genius: Galileo Galilei & His Legacy to Modern Science* (Union Square, 2009).

7 **Enter the young Isaac Newton:** James Gleick, *Isaac Newton* (Harper, 2004).

15 **Long before Newton's discoveries:** Gerrit L. Verschuur, *Hidden Attraction: The History and Mystery of Magnetism* (Oxford University Press, 1996).

17 **It was Benjamin Franklin:** Walter Isaacson, *Benjamin Franklin: An American Life* (Simon & Schuster, 2003).

18 **French physicist Charles Augustin de Coulomb:** C. Stewart Gilmore, *Coulomb and the Evolution of Physics and Engineering in Eighteenth-Century France* (Princeton University Press, 2017).

19 **Frogs?:** Marco Piccolino and Marco Bresadola, *Shocking Frogs: Galvani, Volta, and the Electric Origins of Neuroscience* (Oxford University Press, 2013).

20 **The mythology of science history:** Dan Ch. Christensen, *Hans Christian Ørsted: Reading Nature's Mind* (Oxford University Press, 2013).

22 **Scottish physicist James Clerk Maxwell:** Basil Mahon, *The Man Who Changed Everything: The Life of James Clerk Maxwell* (John Wiley, 2004).

23 **a slow, nonintuitive process:** Ingo Müller, *A History of Thermodynamics: The Doctrine of Energy and Entropy* (Springer, 2007).

25 **Thompson led an adventuresome life:** Sanborn C. Brown, *Benjamin Thompson, Count Rumford* (MIT Press, 1981).

26 **English physicist James Prescott Joule:** Donald S. L. Cardwell, *James Joule: A Biography* (Manchester University Press, 1989).

33 **By one extreme version:** Jeremy England, *All Life Is on Fire: How Thermodynamics Explains the Origins of Living Things* (Basic Books, 2020).

34 **what has been called cosmic evolution:** Eric Chaisson, *Cosmic Evolution: The Rise of Complexity in Nature* (Harvard University Press, 2001).

2. EVOLUTION EVERYWHERE

35 **In its most basic guise:** Eugenie Scott, *Evolution vs. Creationism: An Introduction*, 2nd ed. (University of California Press, 2009). Scott describes more than a dozen different definitions of evolution.

37 **"no matter what you ate . . .":** Mohamed Noor, personal correspondence regarding spoken presentation, April 8, 2025.

38 **"Nothing in biology . . .":** Theodosius Dobzhansky, "Nothing in Biology Makes Sense Except in the Light of Evolution," *American Biology Teacher* 35, no. 3 (1973): 125–29.

38 **what Darwin himself called:** Robert M. Hazen, Shaunna M. Morrison, Sergey V. Krivovichev, and Robert T. Downs, "Lumping and Splitting: Toward a Classification of Mineral Natural Kinds," *American Mineralogist* 107 (2022): 1288–301.

39 **The Evolution of Atoms:** Hendrik Schatz, "The Evolution of Elements and Isotopes," *Elements* 6 (2010): 13–17.

49 **The Evolution of Minerals:** Robert M. Hazen, Dominic Papineau, Wouter Bleeker, et al., "Mineral Evolution," *American Mineralogist* 93 (2008): 1693–720. See also Robert M. Hazen, Shaunna M. Morrison, and Anirudh Prabhu, "The Evolution of Mineral Evolution," in *Celebrating the International Year of Mineralogy*, ed. L. Bindi and G. Cruciani (Springer, 2023), 15–37.

3. SELECTION AND FUNCTION

63 **Scientists call such directionality an "arrow of time":** Richard Morris, *Time's Arrows: Scientific Attitudes Toward Time* (Touchstone, 1986). See also Harold F. Blum, *Time's Arrow and Evolution* (Princeton University Press, 1968); and Stephen Jay Gould, *Time's Arrow, Time's Cycle: Myth and Metaphor in the Discovery of Geological Time* (Harvard University Press, 1988).

65 **On the Nature of Selection:** Michael L. Wong, Carol E. Cleland, Daniel Arend Jr., et al., "On the Roles of Selection and Function in Evolving Systems," *Proceedings of the National Academy of Sciences* 120, e2310223120 (2023).

66 **"What persists, exists":** Blaise Agüera y Arcas, *What Is Life? Evolution as Computation* (MIT Press, 2025).

69 **elaborate dance culture:** Tim Lanman and Edwin Scholes, *Birds of Paradise: Revealing the World's Most Extraordinary Birds* (National Geographic, 2012).

70 **a process called exaptation:** Stephen Jay Gould and Elizabeth S. Vrba, "Exaptation—a Missing Term in the Science of Form," *Paleobiology* 8, no. 1 (Winter 1982): 4–15. See also Steven Johnson, *Where Good Ideas Come From: The Natural History of Innovation* (Riverhead Books, 2010).

75 **These molecules begin to interact:** Stuart Kauffman, *The Origins of Order: Self-Organization and Selection in Evolution* (Oxford University Press, 1993).

4. ENTROPY VERSUS INFORMATION

77 **Some deeply thoughtful and informed scholars:** John Holland, *Emergence: From Chaos to Order* (Helix, 1998). See also Stephen Wolfram, *The Second Law: Resolving the Mystery of the Second Law of Thermodynamics* (Wolfram Media, 2023); and Jeremy England, *All Life Is on Fire: How Thermodynamics Explains the Origins of Living Things* (Basic Books, 2020).

80 **What Is Information?:** James Gleick, *The Information: A History, a Theory, a Flood* (Pantheon, 2011). See also Jimmy Soni and Rob Goodman, *A Mind at Play: How Claude Shannon Invented the Information Age* (Simon & Schuster, 2018).

81 **Perhaps the most basic measure:** Ming Li and Paul Vitanyi, *Kolmogorov Complexity and Its Applications*, 3rd ed. (Springer, 2009).

86 **His one-page paper:** Jack W. Szostak, "Functional Information: Molecular Messages," *Nature* 423 (2003): 689.

88 **A few years later:** Robert Hazen, Patrick L. Griffin, James M. Carothers, and Jack W. Szostak, "Functional Information and the Emergence of Biocomplexity," *Proceedings of the National Academy of Sciences* 104 (2007): 8574–81.

92 **One recurrent idea:** Eric Chaisson, *Cosmic Evolution: The Rise of Complexity in Nature* (Harvard University Press, 2001).

93 **The Law of Increasing Functional Information:** Michael L. Wong et al., "On the Roles of Selection and Function in Evolving Systems," *Proceedings of the National Academy of Sciences* 120, e2310223120 (2023).

5. TESTING THE LAW

96 **The evolution of RNA aptamers:** James M. Carothers, Stephanie C. Oestreich, Jonathan H. Davis, and Jack W. Szostak, "Informational Complexity and Functional Activity of RNA Structures," *Journal of the American Chemical Society* 126 (2004): 5130–37.

99 **Genetic Algorithms:** Mitsuo Gen and Runwei Cheng, *Genetic Algorithms and Engineering Design* (Wiley-Interscience, 1997).

100 **A classic, jaw-dropping example:** John R. Koza, Forrest H. Bennett, David Andre, Martin A. Keane, and Frank Dunlap, "Automated Synthesis of Analog Electrical Circuits by Means of Genetic Programming," *IEEE Transactions on Evolutionary Computing* 1 (1997): 109–28.

101 **NASA's ST5 spacecraft antenna:** Gregory S. Hornby, Al Globus, Derek S. Linden, and Jason D. Lohn, "Automated Antenna Design with Evolutionary Algorithms," paper presented at American Institute of Aeronautics and Astronautics Space conference, San Jose, CA, September 19, 2006.

102 **large language models (or LLMs):** Oswald Campesato, *Large Language Models: An Introduction*, MLI Generative AI Series (Mercury Learning and Information, 2024).

103 **systems that grow simpler over time:** See, for example, Martino E. Malerba, Giulia Ghedini, and Dustin J. Marshall, "Genome Size Affects

Fitness in the Eukaryotic Alga *Dunaliella tertiolecta*," *Current Biology* 30 (2020): 3450–56. See also Aditi Gupta, Thomas LaBar, Miriam Miyagi, and Christoph Adami, "Evolution of Genome Size in Asexual Digital Organisms," *Scientific Reports* 6 (May 2016): 25786, DOI:10.1038/srep25786.

105 **the so-called three-body problem:** Mauri Valtonen, Joanna Anosova, Konstantin Kholshevnikov, Aleksandr Mylläri, Victor Orlov, and Kiyotaka Tanikawa, *The Three-Body Problem from Pythagoras to Hawking* (Springer, 2016).

107 **Nine sequential stages frame Earth's mineral evolution:** Robert M. Hazen and Michael L. Wong, "Open-Ended Versus Bounded Evolution: Mineral Evolution as a Case Study," *PNAS Nexus* 3 (2024): 248.

111 **life is open-ended:** Norman Packard, Mark A. Bedau, Alastair Channon, et al., "An Overview of Open-Ended Evolution: Editorial Introduction to the Open-Ended Evolution II Special Issue," *Artificial Life* 25 (Spring 2019): 93–103. See also Alyssa Adams, Hector Zenil, Paul C. W. Davies, and Sara Imari Walker, "Formal Definitions of Unbounded Evolution and Innovation Reveal Universal Mechanisms for Open-Ended Evolution in Dynamical Systems," *Scientific Reports* 7 (2017): 997.

112 **Molecular Biosignatures:** Marjorie A. Chan, Nancy W. Hinman, Sally L. Potter-McIntyre, et al., "Deciphering Biosignatures in Planetary Contexts," *Astrobiology* 19 (2019): 1075–102.

116 **our method of biosignature detection:** Henderson J. Cleaves II, Grethe Hystad, Anirudh Prabhu, et al., "A Robust, Agnostic Molecular Biosignature Based on Machine Learning," *Proceedings of the National Academy of Sciences* 120 (2023): e2307149120.

116 **A recent example from archaeology:** Yannis Assael, Thea Sommerschield, Brendan Shillingford, et al., "Restoring and Attributing Ancient Texts Using Deep Neural Networks," *Nature* 603 (2022): 280–83.

118 **Agnostic Biosignatures:** Sara I. Walker, William Bains, Leroy Cronin, et al., "Exoplanet Biosignatures: Future Directions," *Astrobiology* 18 (2018): 779–824. See also Stuart Bartlett, Jiazheng Li, Lixiang Gu, et al., "Assessing Planetary Complexity and Potential Agnostic Biosignatures Using Epsilon Machines," *Nature Astronomy* 6 (2022): 387–92.

118 **a long-standing hypothesis:** Steven Benner, "Planets, Minerals and Life's Origin," *Goldschmidt Conference Abstracts* (2013): 686, DOI:10.1180/minmag.2013.077.5.2.

6. WHY HASN'T ANYONE SAID THIS BEFORE?

123 **In a series of influential articles and books:** John Holland, *Emergence: From Chaos to Order* (Helix, 1998); Stuart Kauffman, *At Home in the Universe: The Search for Laws of Self-Organization and Complexity* (Oxford University Press, 1995); Harold J. Morowitz, *The Emergence of Everything: How the World Became Complex* (Oxford University Press, 2002); and Steven Johnson, *Emergence: The Connected Lives of Ants, Brains, Cities, and Software* (Scribner, 2001).

125 **Czech geneticist Gregor Mendel:** Robin M. Henig, *The Monk in the Garden: The Lost and Found Genius of Gregor Mendel, the Father of Genetics* (Houghton Mifflin, 2000).

7. IMPLICATIONS

132 **The German physicist Rudolf Clausius:** Michael Guillen, *Five Equations That Changed the World: The Power and Poetry of Mathematics* (Hyperion, 1995).

134 **"How does life begin?":** David Deamer, *First Life: Discovering the Connections Between Stars, Cells, and How Life Began* (University of California Press, 2011).

135 **Minerals also give organics a lift:** Robert M. Hazen, "Life's Rocky Start," *Scientific American* 284 (2001): 76–85.

137 **Vigorous scientific debates:** Stephen Jay Gould, *Wonderful Life: The Burgess Shale and the Nature of History* (W. W. Norton, 1989); and Simon Conway Morris, *The Crucible of Creation: The Burgess Shale and the Rise of Animals* (Oxford University Press, 1998).

141 **Leading cosmological models:** Chris Impey, *How It Ends: From You to the Universe* (W. W. Norton, 2010).

145 **"first cause" lies at the heart:** Kai Nielsen, *Reason and Practice: A Modern Introduction to Philosophy* (Harper & Row, 1971).

145 **"Nothing is destroyed, nothing is ever lost . . .":** James Prescott Joule, "Memoir of James Prescott Joule," *Memoirs and Proceedings of the Manchester Literary and Philosophical Society* 110 (1863): 1930–31.

149 **"endless forms most wonderful . . .":** Charles Darwin, *On the Origin of Species by Means of Natural Selection* (John Murray, 1859).

Index